MILITARY ELITES

MILITARY ELITES

BY

Roger A. Beaumont

The Bobbs-Merrill Company, Inc.
INDIANAPOLIS / NEW YORK

Our thanks to the following for permission to use the photographs that appear in this book: The Imperial War Museum, United Press International, the United States Army, the United States Army Signal Corps, the United States Air Force, the United States Navy, the United States Naval Institute, the Department of Defense, the Spanish Army, the ECP Armées, Fort D'Ivry.

"The Para's Prayer" from *The French Army* by Paul-Marie de la Gorce, translated by Kenneth Douglas, is reprinted by permission of George Braziller, Inc., copyright © 1963 by George Braziller, Inc.

ISBN 0-672-51977-1
Library of Congress catalog card number 73-22668
Designed by Irving Perkins
Manufactured in the United States of America

First printing

*This volume is dedicated to
my parents
Spencer Alban Beaumont
and
Claire Poser Beaumont
who gave aid and comfort
through many labors and trials*

Contents

	Foreword	ix
I	Elite Units: A Hunger for Heroes	1
II	Children of Frustration: Corps d'Elite in the First World War	15
III	Hitler's Only Victory: Collision of Elites in Spain	30
IV	"Mobs for Jobs": The Decline of Soldierly Honor	44
V	Airborne, Airborne All the Way: The Triumph of Mystique	77
VI	Cybernetic Elites: The Merging of Man and Machines	113
VII	Image and Ideology: Illusions of Elitism	148
VIII	The Ultimate Passion: The Kamikaze Attack Corps	161
IX	The Selection-Destruction Cycle	171
X	Counterpoise to Mass-Man: The Future of Elitism	185
	References	194
	Bibliography	209
	Index	241

Foreword

AS is the case with many "objective" studies, this work has its roots in personal experience. I served two tours of active duty with a branch of the U.S. Army that has been viewed as both elite and marginal at various times: the Military Police Corps. In the process of training and judging men I became increasingly aware of the roles that folklore and selective history—much of it verbal and traditional—play in military life. The tribalism, the totemry, the unproven or false assertions that serve as bedrock in many organizations appear more dramatically in military life because the stakes of the game are so high. Playwrights, historians and novelists have looked at the dynamics of these systems often. It may be that the irrational character of military tradition, a virtual warping of history, is essential in maintaining purpose and cohesion. In any event, the analysis of *corps d'elite* as a phenomenon is overdue. It is amazing how people have accepted the usefulness of corps d'elite without considering their existence as symptomatic or pathological.

This study focuses, therefore, on military elite *units* formed since 1900, not on the generic kinds of elites frequently studied—such as officer corps, castes, cliques, and juntas. The following analysis is not a catalogue of unit genealogies or an encyclopedia of battle imagery. Rather, my purpose has been to synthesize and, by looking at qualitative aspects of the problem, to provide some insights which traditional military history—often aiming at very specific events—fails to do. For those seeking a definitive chronicle of elite units, the bibliography and notes can provide some assistance.

Another goal of this work is to challenge future historians, social scientists, archivists and military and political readers by suggesting how weak is their rational base for discussing or making military policy. Military organizations are poorly understood, especially in view of the fact that they consume such a great proportion of resources. Yet I have relied heavily on the labors of many such organizations, fragmented and partisan as they may have been.

I must specifically thank the following for their encouragement and counsel: Professor Robin Higham, Professor Robert Linder, Professor George Kren, Professor Frank X. Steggert, Professor Martin Edmonds, Professor Morris Janowitz, Professor Bernard J. James, Dean Fred I. Olson, Colonel O.W. Martin, Lieutenant Colonel Burton Eddy, Eleanor Eddy, Captain Karl Farris, Dr. Eugene V. McAndrews, R. Snowden Ficks, Robert Fagaly, Dr. Rainier Hiltermann, Mary Ann Beaumont and Margaret Johnstone.

I would also like to pay tribute to the library staffs of Kansas State University, the Eisenhower Center, the U.S. Armor Center, the U.S. Army Command and General Staff College, Marquette University, the University of Wisconsin at Madison, at Milwaukee and at Oshkosh, the Milwaukee and Oshkosh Public Libraries, the British Ministry of Defence Library, the Imperial War Museum, and the audiovisual branches of the U.S. Navy, Army and Air Force. Also extremely helpful were the military attachés and information personnel of the embassies of the following nations: France, Greece, Belgium, Italy, Portugal, Spain, Switzerland, the United Kingdom and West Germany.

MILITARY ELITES

I

ELITE UNITS:
A HUNGER FOR HEROES

THIS study deals with elite units—*corps d'elite*—formed under military systems in the twentieth century. It is not a catalogue, nor a digest of histories of elite forces. Traditional elites created earlier, like the French Foreign Legion, the Gurkhas, and guard units, are not part of the analysis and discussion, for they were products of different values and societies than those of the twentieth century. Those cases examined illustrate the problems which corps d'elite pose for military and civilian policymakers and analysts—as well as for citizens in countries where military organizations are controlled by elected representation. Furthermore, this study deals with military *units,* not classes or cliques diffused throughout larger organizations in key roles, like the German general staff, the graduates of West Point, Annapolis or Sandhurst, or the survivors of the Long March.

Three main questions are posed about the elite forces formed since 1900. First, why did these units thrive in the face of collectiv-

1

ization? How did they reflect or contradict the values of their parent systems? And how much did corps d'elite match their creators' hopes and justify immunity from orthodox control and special access to resources?

Certainly it would have been reasonable for someone appraising trends in 1900 to have predicted a decline in the forming of elite units in the twentieth century. After all, the late nineteenth century saw an advance of mass production, mass education, mass media, mass marketing and mass politics throughout the world. In Europe, the center of world power, mass armies, and navies as well, based on conscription, were becoming a normal form of national defense. True, there were many elite forces in existence, glittering ceremonial remnants of the past. But it was clear after the American Civil War (1861-65) and the Franco-Prussian War (1870) that the brunt of battle in the machine age would be borne by masses of conscripts led by highly trained officer cadres, carried to war by railroads and controlled by telegraph. Of course, the best minds were not terribly concerned with war. The new century opened on a note of social progress, in medicine, law, politics, industry, government and education. With all these mighty forces shaping what Ortega y Gasset later called "mass man," the remaining corps d'elite faintly resembled dinosaurs, colorful remnants of a different world.

It is, of course, essential to consider what an elite unit *is*. The elite forces formed in the twentieth century were, naturally enough, voluntary and, with rare exceptions, had official recognition of their elite status before they were actually assembled. Usually, this meant that physical and mental standards for admission were high. They had distinctive uniforms or insignia and traditions and customs—frequently highly synthetic—which formed rapidly. The new elite forces also were relatively free from

2

ordinary administration and discipline. Entrance to these units was often through the surviving of an ordeal, a "rite of passage," requiring tolerance of pain or danger and subsequent dedication to a hazardous role. Many corps d'elite—but not all—received much publicity from media. Moreover, they did not have to accept nonvolunteers into their company. These factors produced in many cases a disdain among members of elite units for outsiders, which meant that headquarters personnel, civilians, and adjacent allied units were often more detested than the enemy. Elite forces, however, that had mental or technical functions showed less inclination to brawling, more cooperativeness and, generally, less adolescent unease. Across the many types of elite forces the most common traits have been voluntarism, special selection criteria and training, and distinctive clothing or insignia.

A review of the types of elite forces is also useful. At the same time, it should be noted that the basic purpose for creating the organization was usually changed by the pressure of events.

Types	Example
Ceremonial/Social	Philadelphia Black Horse Troop
Combat proven	The Iron Brigade
Praetorian (bodyguard)	The early SS
Ethnic/Cultural	Gurkhas
Politico-ideological	International Brigades
Romanticist/Atavistic	1st Special Service Force
Technological/Proto-Cybernetic	79th Armoured Division
Nihilistic	Spanish Foreign Legion
Functional/Objective	Aviators

The elites in existence at the beginning of the twentieth century were of the first three types, which, formed after 1900, are generally found toward the bottom of the list. Some units, like the

3

British Brigade of Guards, came to be a mix of types. And some shifted from time to time, like the better combat units of the *Waffen* SS.

Modern military elite units have been children of the storm, products of crisis and instability. The implication of their creation is often a response to enemy success or threat of advantage. In such a "fire brigade" role, they trade youthful daring for time, trying to make up for deficiencies in military hardware revealed in the early phases of hostilities. In another sense, elites have often been the mirror image of successful enemy units of the same ilk. The most rational form of corps d'elite were those units formed to operate a technological system. Paradoxically, with only a few exceptions, membership in an elite force has been the only reward in a career sense. While U.S. and French airborne veterans found their affiliation useful during 1945-60, there has not been, overall, a strong correlation between elite force membership and promotion in the general structure of the parent armed force.

In a similarly paradoxical vein, the creation of corps d'elite often failed to solve problems and sometimes created new ones. Their special access to status and resources produced intra-organizational tensions, a kind of military class war. The image of these forces often obscured less pleasant side effects, and military history too often was written in combative cliches, with loving attention paid to the movement of units and the clash of personalities, without concern for more real and quantitative problems.[1] Thus the decision to create such forces was often made impulsively or intuitively. With tiresome and tragic regularity, forces raised to cope with a specific problem often found the original challenge gone by the time they were ready for action. This led to their being used—and misused—for other purposes. Among ground forces, corps d'elite were almost always lightly equipped. Their role was usually that of hit-and-run raiders.

Nevertheless, their zeal and the high quality of the personnel, as well as a shortage of manpower in the parent forces, led to their being employed in prolonged fighting alongside and against standard units who had heavy weapons and better support facilities. The most serious question is that of "leadership drain." How much did special units rob their parent organizations of leadership? Should not the privates in the U.S. Marine Corps, the air forces, or the Royal Marine Commandos in World War II have been sergeants and lieutenants in the conscript army? The effects of clustering together the eager and fit and putting them into risky and intense battle is discussed further in the chapter on the "selection-destruction cycle."

The extent to which military elite units became victims of their own posturing and publicity underlines their role as tuning forks of their parent cultures' values. The hunger for heroes is perhaps greater today than it has ever been, in our "Age of Anxiety" where men "lead lives of quiet desperation," torn from the soil; victims of blind forces and heedless cruelties; debased physically and mentally by repetitive and meaningless work, synthetic food; spectators of tawdry thrills; fearful of death in an age without gods. The mythic power of the elite force must then be considered as a symbol of individual courage triumphant in the face of the machinery of destruction. This hunger became manifest in the cinema and the comic strip, where crude popular heroes abounded. Thus, the compulsive groping for romantic heroism seen in the literature and the arts of the nineteenth century was debased and extended in the age of mass communication. The myth of the warrior hero, after all, is ancient and universal and has not faded with time. The warriors and warrior-kings of the Old Testament, the Icelandic and Norse sagas, the *Iliad,* the *Nibelungenlied,* the Anglo-Saxon martial adventurer tradition, Roland, Chaka, Saladin, Bayard, and the 47 *ronin* are brothers under their armor. That

5

array of martial heroes around the world underlines the fact that viewing corps d'elite as mere tactical instruments misses their role as a symbol of aggressiveness. Corps d'elite tell us more about their parent societies than their creators meant to. And this is especially significant since elite forces flourished around the world in the twentieth century, appearing again and again at key junctures in history in an even more active and visible way than before. The urbanizing of Western man may also have made some difference insomuch as the manpower base of sturdy, pious folk, used to living outdoors, with good nerves and digestion, has been vastly reduced, as it was in the late days of the Roman Republic. Half of the youths examined for the draft in the United States during the peak of the Vietnamese War were rejected on physical grounds. By mid-century, only totalitarian states like Russia, China, Spain and some neutrals—Sweden and Switzerland—were resorting to universal military service. The uneven performance of American conscript infantry in Korea and Vietnam, their dependence on creature comforts and technology, and their political volatility drove home the fact that the Communist powers' strategy of Wars of National Liberation hit the Western allies where it hurt. After 1945, brushfire war in Palestine, Cyprus, Ireland, Algeria or Indo-China saw corps d'elite responding to blazes set by terrorist cadres.

Behind the question as to why the elite forces are created by higher authority is another question: Why are men attracted to join them? Is it privilege or status? Is it extra pay? Or is it a search for meaning in an age of facelessness? The men who survived the screening and the ordeals of entry into corps d'elite and into battle gained a greatly enhanced sense of importance and achievement. They felt tough, heedless of death, favored by fate, and sure that they were many cuts above the average man. The increased sense of power that comes to those who survive great enterprise and danger is a common theme in history. That Harmhab, the tough general, overthrew Ikhnaton, the peaceful, dreaming pharaoh, and that the Praetorian Guard ruled Imperial Rome are cliches. That young men love rough sports, danger and the companionship of their fellows is hardly a revelation. Therefore, the seeds of unit elitism can grow many wildflowers. Yet shapers of policy and military commanders regularly ignored the powerful potential for dangerous side-effects set loose by the creation of corps d'elite. Did John F. Kennedy in the early 1960s, in expanding the Green Berets, reach for a straw in the political wind, or did he see the Special Forces as a possible panacea in the Vietnam tangle?

For twentieth-century military elite units have a darker side to their nature. The use of "special forces" is a virtual acceptance by their masters that they are outside the laws of war and the limits of civility. Most modern corps d'elite were expected to behave more like gangsters than soldiers. The normal tactics of submariners in the twentieth century would have been seen as worse than piracy in the nineteenth. Not only was this extralegal bent the case in totalitarian states, Left and Right, but in Western democratic nations a similar swagger and flagrant contempt for the rules of the game appeared. Such organizations as the U.S. Rangers, the British Commandos, the Special Air Service, the Long Range Desert Group, the U.S. Marine Raiders were trained to kill stealthily, to maim cruelly, to attack men unaware. The main targets of such forces and of the airborne were not enemy combat soldiers or resolution by battle, but harassment of third-rate troops and civilians in rear areas. Like guerrillas, these elites were to avoid anything resembling a fair fight. The enthusiasm which the exploits of these units fanned in the publics of their parent nations in war and peace is grim testimony to the human fragility of the traditions of civility under stress.

The frequency with which elite forces of many nations indulged in excesses also suggests how these units resembled a primitive warrior band. The recurrence of tribal symbols and images resem-

7

bles the same grim esthetic, that which leads small boys to build models of aircraft and warships and to fondle weapons. Powerful emotions are generated by indulging young men in their appetite for military totemry.[2] The refusal of airborne enlisted men to salute "leg" (non-airborne-qualified) officers is an example of this effect. The ambush staged by a special operations group to frighten away headquarters representatives described in Tom Chamales's *Never So Few* is another. In the Second World War, higher commanders so resented the prima donna behavior of elite forces that many were disbanded during and after the war. At odds on many questions, Field Marshal Montgomery and General Omar Bradley shared a disdain for special units.

Yet, in spite of all these problems, it is an *advantage* perceived by those in power that leads to the creation of corps d'elite. In both World Wars and in Spain, special units were formed in the face of retreat, defeat or frustration. In that sense they are more symptomatic of the state of mind of commanders and leaders under pressure.

Yet in less stressful moments, why has so little thought been given to what is a critical indicator, a basic design question in military organization? The danger of elitism in a broader sense has, after all, long been a concern to the supporters of democracy. Elite castes and classes were among the first topics studied by behavioral scientists in the late nineteenth century when Marx's hypothesis of class war and Darwin's version of diverging fans of evolution underlined the sense of rank and difference in society.[3] Many philosophers, scientists and politicians defended the distribution of power within societies and governments on a pseudo-scientific basis.

The fears of early American populists of corps d'elite as seedbeds of reactionism were well-founded. Washington drew the teeth of an elitist mutiny among Continental army officers in the affair of the Newburgh Addresses. Nevertheless, in the twentieth century, elite units became a symbol of aggressive, decisive, personal action, in contrast to the vast noncombatant support bureaucracy of modern mass armies and navies. In the 1950s, the pictures of cliff-scaling Rangers and mass parachute drops during Cold War maneuvers were anachronistic. But, like cavalry charges a generation earlier, they served as visual spice. And, of course, casualties among volunteers did not create the political feedback they might have among conscripts. The hunger for decisive action among literate publics was fed by such adventures as the RAF Bomber Command's 1940 attacks on German cities and General Doolittle's raid on Tokyo in 1942.

Our reasonable man pondering things over on New Year's Eve, 1899, could have logically expected new corps d'elite, if any were to appear, to be less romantic than, say, the Brigade of Guards, and rather more along the lines of ships' engineers who, however vital, could not share the lofty status of Navy line officers. Many new specialties led to the creation of staff jobs. Even in the nineteenth century, auxiliary units for servicing of technology abounded, such as signal and railway corps and expanded medical services. But the twentieth-century corps d'elite did not merely supplement fighting services but actually upstaged them, although many such forces were created with technological change in mind. Glamour came from mystique rather than technique.

The twentieth century also saw the shattering of the social stability of the past. The aristocratic model had been crippled in the eighteenth century in France by the guillotine and in America by the musket, and in Britain in the nineteenth by the ballot. In the twentieth century the remnants were swept out of power almost completely. Nation after nation came under the rule of new leaders, most of whom were far from the athletic, eccentric nobility. Rather, the virtues of the bourgeois triumphed in the democratic West and in the autocratic East. The business-suited bureaucrat, thick in the middle, became a symbol of Communism, just as the business-suited politician, manager and bureaucrat—military as well as civilian—thick in the middle, became the symbol of authority in the West. Soldiers and sailors wore mufti in Washington and London to minimize their visibility in peacetime years of budget trimming. The public hunger for heroes was sated in some countries by a reversion to posturing strong-man dictators. Western industrial countries turned to the screen and popular literature for fantasy leader images, decisive, personally courageous, forthright, handsome, physically vigorous men and women, types notably lacking in the world of politics and business. The military bureaucracies saw a similar change in the upper levels, and the corps d'elite in the military world became an escape for the young and adventurous. Military organization did not generally reflect the extent to which young men were more and more actually fighting the wars, and elite forces were a good place to escape the humdrum, desk-bound, careerist, harem-like world of bureaupolitics, the committee-ridden anonymity, consensus and old-boy petty politics that passed for a command structure. Corps d'elite were one way to keep the Young Turks' attention away from the stodgy and too often dangerously stupid upper layers.

Along with such pressures toward conformity and averaging in military command, the model of the future at the turn of the century seemed to be the mass conscript army, mobilized and controlled by an intellectual general staff through sophisticated use of telegraph and railway. Using such a model, the Prussian general staff had seized the tiller of German history in 1859. Even those proverbial military muddlers-through, the British and the Americans, adopted general staff models as the new century began. Yet the experiences in slaughter and futility with the mass

10

army in the American Civil War had been overlooked. The colorful campaigns of Stonewall Jackson in the Mississippi Valley and
Lee at Chancellorsville dazzled European students before 1914.
They ignored the decisive but dull trench warfare at Petersburg
and the strategy of total war through blockade and devastation
in the Shenandoah Valley and Georgia. Thus the lessons dearly
won during 1861-65 were learned again at Mukden and Port
Arthur and in the grisly futility of 1914-18. In the last year of
World War I, the successes of tanks and aircraft and the spectacular if vain breakthroughs of German storm troops reflected the
bankruptcy of the big battalion system. By 1917-18, head-down
battering tactics produced large-scale mutiny among the survivors
in Russia, Italy, France, Austria, and Turkey.

From 1916 on, guerrilla warfare assumed new prominence as
a gadfly response to the sluggish reflexes of mass forces. In
Arabia-Palestine, Ireland, and East Africa, tiny bands tied up vast
armies. In the 1920s the Spaniards and French were bedeviled by
the Rifs in Morocco and Algeria, and Ireland became free. In the
1920s and '30s the Russian Civil War, the Syrian Revolt, Spain,
and the Gran Chaco War highlighted guerrillas. The Second
World War weakened the colonial powers by revealing their military flaws and by providing sophisticating experiences and guns
to subject peoples. Elite forces of the colonial powers became a
standard response to the resulting insurgency. Although often
functioning as virtual police, the basic ethic of such corps d'elite
continued to be personalized combat in conventional war. Sending
such forces against guerrillas often led, not surprisingly, to frustration, the father of atrocity. Indeed, the urge of elite forces for a
stand-up fight became a point of leverage for terrorist-guerrilla
movements. The zeal and the combativeness of the corps d'elite
had only to be tantalized and confounded over time. Eventually,
units would be goaded into excess, thus gaining support for the

11

terrorists. That much of the world expected a double standard in behavior in "limited wars" eluded even the most perceptive spectators of the wars of "national liberation."

The formal sponsorship of the Communist world of such shadow wars in the middle 1950s came in the wake of American victory in Korea—one million Chinese and North Korean dead versus 33,000 Americans—stalemate and partition in Indo-China, and Communist defeat in Malaya and Greece. These costly confrontations with Western military technology underlined the lesson learned by Japan during 1941-45. The new model would be cheaper and slower, but it took into account the pre-eminent role of artillery in "conventional" warfare. To the average man, infantry appeared as the main element in modern battle. But this view overlooked the fact that artillery did about eighty percent of the butcher's work in both world wars. Infantry had been drifting closer to marginal utility since the first major victory of rifles in the defense at New Orleans. The Confederates' Enfields, the Boers' Mausers, and the Kaiser's machine guns drove home the lesson time and time again. At Verdun in 1916, the crescendo of artillery made the infantry mere ants on an anvil under the blows of giant hammers. The power of massed indirect artillery fire and machine guns forced a dispersal of forces in the field. And air power amplified the effect. Yet the realities of this dispersal have been remarkably slow in sinking in, in military organizations as well as in the minds of civilians.

The role of infantry as staked goat with artillery as the hunter was reflected dramatically in U.S. Army military historians' battlefield interviews in World War II and Korea. They found that "less than twenty-five percent of men in combat . . . fired their weapons at all" and that only fifteen percent participated in fire fights on a fairly regular basis.[4] Yet this was less upsetting to the self-image of commanders and nations than might have been expected. Bureaucracies and professionals cope with unpleasant truth by ignoring it or by rationalizing. Marshall's observations were merely ignored. Although his data were available soon after the war, it was not until the late 1950s that American training systems addressed this problem of low participation. Yet the Soviets, whose artillery techniques were cruder than the Western powers', nevertheless awarded special status to that branch as "the god of war." The celebration of Artillery Week and the status of Chief of Artillery as a fully ranked service head displayed their esteem for the principal combat arm.[5]

As artillery grew in importance, military technology also produced more and more noncombatants among those armies organized around the new systems. Ratios of men "back" to men "up" tilted in favor of support forces to the point where estimates of service versus combat elements in U.S. forces compared with combat troops by the end of World War II were in excess of ten to one. In the late 1960s in Vietnam, 600,000 American troops were committed to support 70,000 combatants. At the same time, the relative role of the elite component within the shrinking percentage of fighters became larger.

Unfortunately for our rationally minded prophet of 1900, the impact of standardized industrial technology and other collectivizing pressures was short-lived. The First World War brought these pressures to a head in a great contest. The result was stagnation, stalemate, a war of "attrition," of grinding down. Massed artillery, machine guns, and barbed wire created conditions of primitive siegecraft that Caesar would have found familiar. The response to the problem was a return to elitism, and new aristocracies of technology and of shock troops emerged. The new pattern of war resembled a medieval tournament and persisted into the Second World War and after: many would watch and support. The few—often, as in the Battle of Britain, the very few—would

13

fight. Most men in uniform would become drab clerks, teamsters or freight handlers. The average line-infantry units consisted mainly of men without the personal skills, social base or resources to escape.

It is in the garden of the trenches of World War I, however, that we must look to see the first shoots of the unexpectedly fertile and wild new species of corps d'elite that evolved with their roots richly watered with the blood of repeated frustration.

II

CHILDREN OF FRUSTRATION:
CORPS D'ELITE
IN THE FIRST WORLD WAR

FIRST, a visit to a dugout on the front line in late 1917. The men are huddled around candles. Their faces are gaunt and drawn. They are about to attack. Some are staring into space, some are praying, some are sleeping, or appear to be. The candle flames flicker and the men shift uneasily as the guns thunder in a rolling cadence. The head of an officer appears through the blankets that form the gas lock. "One minute. Stand to." The men rise sluggishly and shuffle out, some dragging their weapons. They will be going over the top, into the enemy wire and machine-gun belts, blizzarded by enemy artillery, mortars, rifles, grenades. In two hours, only one in three will be alive. These are human beings who are being used to batter through the enemy front. Their own forces' barrage has given away the hope of surprise. The land over which they are to advance has been chewed into a swamp. They know that their protective barrage may fall short or creep ahead of their attack and make them easy targets. They are men under

15

sentence, conscripts of all ages from eighteen to forty, thin and fat, reluctantly marching to execution.

Then, let us visit another dugout in another sector. Here is another group of men. But these men are talking animatedly. These men are young, lean, bright-eyed, quick-moving, athletic. Jokes are being told, maps being studied. These men carry knives, light machine guns, pistols, grenades, rather than the heavy packs and rifles of the other group. Some have brass knuckles. They wear skull insignia on their uniforms. Two are sharpening their knives in smooth circular motions on a whetstone. There is no barrage rolling outside. But these men, too, will soon be going to the attack, not as sheep driven over a cliff but as individuals filtering through the enemy's defenses, killing rather than being killed. There will be only a short barrage to cover their crossing no-man's-land, firing smoke and gas to confuse the enemy. These men will be not a bloody battering ram but bacteria which infect the enemy organism with the disease of defeat, slicing at nerves and arteries, at headquarters, supply dumps, bridges, phone lines. These are the pick of the forces. These are shock troops.

The holocaust of the First World War still staggers the imagination, and the scale of slaughter and agony oppresses the mind. No one expected it. The one heartening aspect of that conflict is the extent to which experts were wrong, although the example does not seem to have had much impact. The most frightening part was the slow-witted performance of most military and political leaders in recognizing what was happening at the time. The armies and the weapons of 1914 had been designed for a fast series of clashes, with an end to hostilities in weeks, or at most, a few months. The French were seeking *revanche,* revenge for 1870. The Germans hoped for a quick victory in the west and then in the east against Russia. The British hoped to balance power. The game plan was basically an old one, but there were some new plays. On land and sea, there was a lack of quick-wittedness and of communications and information systems, which could control the massive forces that had been set loose. In battle after battle, the high command was one or two jumps behind developments. Their orders—obeyed by men on the spot in a spirit of patriotism that eventually went sour—often arrived with the situation vastly changed. The ability of the headquarters to perceive and command was minimal once the forces were committed. But trappings of rank and power remained, as did habits of an age of simpler war and the myth that the people at the top were in control, that somehow they knew more.

The series of continuing disasters was complicated by the misunderstood momentum of mass production. All the great armies in the end had to abdicate much of their authority to civilian planners and producers. Senior commanders became surrounded by staffs of civilians in uniform and by technicians and scientists whose work they barely understood. The problems were too many and too complicated for any one man to grasp. The collision of these blind forces set up new patterns of shock waves that led the war further and further toward what every war carries in its wake, the Fifth Horseman of the Apocalypse—Revolution.

The actual flow of events was hard for the publics in the major industrial nations to see. They were bombarded with posters and slogans, lied to through a censored press. The complexities of command, priorities and procurement, of weapons development lead-time were beyond the grasp of many generals and politicians, let alone civilians. The only real indicators were the decline in the standard of living and of social values, the trainloads of wounded, the casualty lists and the nervous or withdrawn behavior of the men on leave. Like the American Civil War, as the burden of the war and its length became unbearable, the drive for blind victory became more and more obsessive.

Time and time again, as the war dragged on, each side tried

17

to "break through." The Germans were relatively successful in the east, where the primitively led and equipped Russian army was gradually stretched and worn out. But in the west, millions of men had burrowed into the earth in late 1914. The subsequent attempts of each side to batter through the other's trench and barbed-wire systems drenched the soil of France and Flanders with showers of blood for the next four years. The massing of thousands of heavy guns, the use of poison gas and flamethrowers and even the tanks in 1916 failed to do more than shift the line slightly. Losses of eighty percent in infantry units became common. Sixty thousand British troops fell in the first three hours on the Somme. The rail systems behind the front, out of artillery range, allowed the easy shunting of forces to plug a threatened gap with reserves. The French Army was bled white at Verdun, and the British reluctantly began conscription in 1916. The guns thundered steadily, the machine guns chattered, the wire forests thickened. By the hundreds of thousands men went over the top, struggled forward, were shattered, blasted, maimed, deformed, their minds broken. Battalions of physicians were swallowed up in the charnel house trying to stanch the collective wound.

At the same time, the complexity of technological war and the conscription system worked at cross purposes, especially in a protracted war. The discipline of the various armies was harsh, but useful in a crude sense in maintaining order among reluctant soldiers in a short conflict. The French Army in particular, but the Germans, Russians, British, Turks and Italians as well, had problems. The growing literacy and political sensitivity among enlisted men caused resentment toward what seemed to be feudal privileges. As it became more and more apparent that the old way was not working, changes began to appear. The frustration of military deadlock became the source of social revolution. It was very difficult to label the elitist movements in the various armies

18

as right or left wing. The most dramatic cases were those of the German storm troops and the Italian *arditi*. In both instances, new forms appeared which transcended the old social order and values, and produced a classless band of heroes who defied the all-devouring technology of war.

Most of the major armies' discipline and cohesion was eroded as time went on. The French army mutinied. The Russians collapsed. The Italian junior officers revitalized the army. The German Navy mutinied at Kiel. Disaffection at the front and postwar disturbances created a need for the burden of fighting and offensive action to be carried by a small, highly motivated minority, e.g., the Australians, German machine gunners, the Alpini and *arditi,* the Battalions of Death of the Kerensky period, and tank units of the British and French forces. Indeed, the aggressiveness of American units was embarrassingly obvious, even to the Australians.

For at the lower levels, where the price was paid, men were thinking. Captain Liddell Hart, wounded on the Somme, was thinking. Corporal Hitler was thinking. Sergeant Mussolini was thinking. In Russia, thinkers moved to revolution. In the German Navy, mutiny was the end result. Revolutions in tactics were also born out of the stalemate. They took various forms. The victory of massed armor by the British Tank Corps at Cambrai in October 1917 was a glimpse of the future, but the British Expeditionary Force HQ was not prepared for such success and there was no follow-up force ready to exploit the easy penetration. It would, however, be infantry corps d'elite that played a major role in the breakthroughs and advances carried out by both sides in the final year of the war. The Germans struck first.

The storm troops who served as the spear-point of German offensives in 1917 and 1918 were the product of modern quantitative thinking. Their development resembled the review of a prob-

19

lem in industrial engineering. In the storm battalions, modern methods of selection and training were blended with special weapons into a system reflecting a spatial and technical view beyond the reach of Cromwell, de Saxe, or even Napoleon.

The use of this carefully honed and forged technique in the greatest gamble of the war was an attempt to resolve Germany's dilemma in the winter of 1917-18. The big naval battle at Jutland in May 1916, between the British and Germans, showed that the Allied blockade could not be broken by the Kaiser's High Seas Fleet. The near-victory of Germany's U-boats was reversed when Lloyd George, Britain's Prime Minister, ordered the Royal Navy into the business of convoying. The starved and battered Turks were reeling back in Palestine. The Italians, after a seeming collapse at Caporetto, rallied as the young officers of their army waged a revolution in the trenches. And the Americans were on the way. The only bright spot on Germany's horizon was the removal of Russia from the Allied ranks. Although Bolsheviks and other groups nibbled at the German eastern occupation forces, the peace of Brest-Litovsk between the Germans and the Leninists freed eighty German divisions for use in the West in the spring of 1918.

The German spring offensive of 1918 with its Wagnerian code names had a mythical quality. The first waves of the attack followed new tactical concepts already tested successfully against the Russians at Riga and against the Italians at Caporetto in 1917. The new system of infiltration rekindled the individual heroism, for instead of the brute collision of massed battalions in a *Stahlgewitter*—steel storm—infiltration tactics resembled scalpels thrusting through the skin of the enemy "body," to cut vital nerves, sinews, and organs.

Like many radical tactical concepts, infiltration was the product of relatively junior officers. While General von Hutier's name

20

was linked to the system subsequently,[1] it was a synthesis of the artillery systems of Colonel Bruchmüller, an artillery commander (whose activities with a traveling "circus" of railborne guns led to his receiving the nickname of "Durchbruchmüller"), and Staff Captain Geyer, also an artilleryman, who first articulated his own system in *The Attack in Trench Warfare*. (Meanwhile, Captain Laffargue, a French infantryman, wrote along similar lines. At their best, French infantry tactics became more deft and sparing of life than those of the British.)[2] Junior infantry officers had also been experimenting. In late 1915, a Captain Rohr and a Major Reddemann led *Stosstruppen* into action near Muhlhausen.[3] These units, composed of volunteers under twenty-five, placed a high premium on hand-to-hand fighting and had a strong inclination toward brutality. They adopted the death's head as their insignia—as the SS and Gestapo did a few years later—and were provided extra rations. There was a notable reduction in rank consciousness and such signs of comradeship as enlisted men addressing officers with the familiar *du*. These individual experiments were eventually orchestrated and became the backbone of the last great German offensives of the war. Von Hutier directed the application of the system in the attack at Riga, and his XVIIIth Army was slated for a key role in the spring attack series against the French on the Chemin des Dames. Infiltration, however, was to be most widely used in the spring of 1918.

The designers of infiltration tactics visualized trench warfare from a higher altitude than had the advocates of slugging attacks. The goals of the infiltrators were weak points in the line and then the nerves and arteries of the enemy armies, such as headquarters, crossroads, and supply and communication centers. Reliable, aggressive, intelligent troops were needed to carry out the individual assaults. The attackers in 1918 were aided by the fact that as the war wore on, the losses and the recognition by headquarters of the overwhelming role of artillery led to abandonment of the front-line positions. Instead of manning long ribbons of trenches with troops, widely spaced listening posts and strong points monitored the front, with the bulk of forces held back to respond to attacks in coordination with artillery.

The ranks of the German forces in the west were culled for volunteers to provide manpower for the storm battalions, who were then trained in independent action and the use of light automatic weapons and flamethrowers. Emphasis was placed on using initiative and bypassing strongpoints. The storm troops led other units through the breaks in the enemy front. Commanders were to avoid pausing to reassemble but rather to maintain momentum. The artillery preparation program for infiltration omitted preregistration firing that usually warned the enemy before an impending attack. Instead, the fire plan called for a short, sharp barrage against critical road junctions, headquarters, and opposing batteries. Gas and smoke rather than high-explosive shells were used in the immediate pre-attack bombardment against frontline positions. This avoided the churning up of ground that had impeded so many attacks. The first wave of the infantry—the storm troops—then isolated and bypassed strong points. These were left to be reduced by the second phase of the storm attack with light artillery.[4]

The result of these German spring assaults of 1918 was a tragicomedy. The blockade won a subtle victory as the Germans, long starved of creature comforts, often became looting mobs, diverted by the liquor and cigarettes of the British dumps. Lengthening flanks on the bulge formed by the attack required more men and produced higher casualties than originally projected. By mid-April, the attack bogged down in the British Fifth Army area. Leaders had been sacrificed to no purpose. From late March to

June the Germans thudded away at the Allied gates, and they gained much ground, but in the end, failing to break through, they faced the same array of impending crises as before. With the loss of its leadership cadres, the German Army began to lose cohesion during the summer. The loss of the flower of the forces without victory hurried the collapse of the German Army in the West. Morale ebbed, and small unit commanders began to falsify patrol and position reports. The spirit of the *Sturmabteilung* (SA), however, the forlorn hope of Imperial Germany, haunted the next generation of Europe and the world. The images of *Frontgenossen* (front comrades), the cult of the *Schutzengraven* (trenches), and the *Sturmabteilung* (storm troops) were used by the Nazis as symbols of ruthless force in a victorious antidemocratic revolution.

Just as linkages between the *Freikorps* and the early SS and between the storm battalions and the SA can be traced, there was also a connection between the storm battalions of 1918 and the *Waffen* SS through the efforts of SS General Felix Steiner. A storm-battalion veteran and regular officer, Steiner clashed with the traditional leaders of the *Reichswehr* who eagerly reverted to infantry-artillery tactics after 1918, rejecting infiltration and armor. Steiner proposed new tactical systems, techniques, and elites that were as revolutionary as the political changes proposed by extremists of the Right or Left. De Gaulle, Fuller, and others glimpsed a new age in which the tactical systems they postulated would force broader changes. Some were more interested in the tactical problems themselves, and no doubt most sensed that a way to advancement lay in undercutting the tribal elders' monopoly over the forms of power.

Steiner rejected parade ground drill, stressed sports, and reduced distinctions of rank. As a result, relations between officers and men in the SS (and generally in the German Army) in World War II were more personal and "democratic" than in the United States forces.[5]

But the storm troop concept, when confronted with the barriers of mass, material, and space, failed to produce an advantage. While perhaps the odds were too great, Steiner's system was based on a much defter and less frequent use of elite forces than proved the case in the field. In any case, Steiner also aimed at making the SS a repudiation of an archaic system, as well as an overshadowing of the German Army with a National Socialist rival. The uneven expansion of the SS in the Second World War stretched and tore the fabric of Steiner's system, so that few of its units attained the quality that he foresaw.

The effect of the storm troops was also felt by the German Army's opponents, particularly in Italy. There, following the infiltration successes at Caporetto in October 1917, the shattered Italian Army underwent a spontaneous purging. One of the symptoms of revitalization was the creation of a volunteer corps of young shock troops, the *arditi*—the "brave ones." (The *arditi* of the Army and the early Fascist era were not the *arditi del popolo,* left-wing paramilitary street fighters who successfully defended Parma against the *squadristi* in 1920.)[6] Like the *Stosstruppen,* their training emphasized close personal combat and the raid, using knives and hand grenades. A high percentage of the volunteers were from the urban middle class, with university backgrounds.[7] Costanzo Ciano (father of Galeazzo, later Mussolini's foreign minister) was their commander. The *arditi,* who dressed in black shirts and distinctive headgear and who adopted a nihilistic motto of "Who cares?" (politely translated), soon became recognized as a corps d'elite.

It is hard to see where the *arditi* ended and the *Fascisti* began. Mussolini, after his discharge from the army, festooned his editorial office with *arditi* banners, grenades, helmets, and daggers as he took up the cause of the resentful veterans:

> The ex-servicemen are returning . . . in twos and threes. They haven't even the esthetic satisfaction of seeing themselves received triumphantly. . . . The soldier who returns with the intimate satisfaction of having done his duty which permits him to look upon those who forgot their duty looks for work and there is no work.[8]

The *trincerismo*—trench-mindedness—of the *arditi* was suggestive of the Black and Tans, *Freikorps, Stahlhelm,* and *Sturmabteilungen,* as well as the surrealistic images of Frederick Manning and Ernst Jünger. The *arditi* song, *"Giovinezza"*—"Youth"—passed into the Fascist hymnal. Their members went with d'Annunzio to Fiume in 1919 and became the core of the original Fascist party at San Sepulcro. Ultimately they were linked with the gorier excesses of the Fascists, including the Matteotti murder in 1924.[9] In July 1920, *squadristi* were formed in the Fascist party in response to *arditi* defection, signaling the demise of the *arditi,*[10] although the blackshirt military units of the Italian Army later retained *arditi* "folklore and choreography,"[11] if not their military effectiveness.

Bullying and brutality, e.g., the castor-oil treatment, alienated the more idealistic *arditi* and created a stereotype of the Fascist regime in the eyes of much of the world as gangsterlike. Its origins in defeat were ironically prophetic of the grotesque overkill in the Italian invasion of Ethiopia, the Spanish debacle, and the rout of (for the most part) indifferent and poorly led, trained, and equipped Italian forces in Spain, Albania, Greece, North Africa, and Russia. In a regime built on publicity and posturing and devoid of cohesion, only a few units, some elite regiments of the Army, the sailors of the 10th MAS and individual *Regia*

Aeronautica pilots, offset general physical and organizational bankruptcy. Less measurable was the impact that the image of blackshirted Fascism in Italy had on the forming of the black-uniformed SS "elite" units by the Nazis. As was so often the case, the power of the corps d'elite as a symbol exceeded its physical capacity.

The more traditional elite forces, such as the British Guards regiments, the French Foreign Legion, and the U.S. Marine Corps, fitted into the orthodox mold of the 1914-18 war. They were brave and reliable: they were not the source of effective counter-strategies to the bloody debacle. The danger in new solutions to the tactical quandary was that they were not merely mechanical. Changing the nature of armies, of soldiers, and the roles had inherent dangers. Sanctioning brutality, lowering class boundaries, reallocating status and resources, changing the age profile of command were models for change in the society. The Tsars, the Hohenzollerns, Hitler, Mussolini, and the Soviets knew the power of armies as levers for influencing civil society. The swiftness with which the "old guard" returned to pre-1914 military systems and traditions of soldiering in America, France, Britain, Italy and Germany has been derided by those who saw the rejection of armor, aviation and shock troops as blindness to military necessity. But the guardians of the old way knew the new weapons were a young man's game, and that more than tactical advantage was implicit in their use. Real reform, they sensed, would put too much power in the hands of junior officers, as it had in Turkey and Italy. And so from 1918 until the middle of World War II, the battleship dominated naval tactical thought in all major fleets. Young George Patton and Dwight Eisenhower were warned that their enthusiasm for armor would hurt their careers. It was significant that the Soviet Union led the way in the 1920s in innovative tactical developments, and that the reac-

26

tionary Stalin eventually saw fit to decapitate his army. For military reform does not stop at the boundaries of armies or navies, but also alters the nature of their parent societies.

The Allies in the First World War also had many types of elite forces, most of them traditional, but also new forms, like the French Army *Nettoyeurs*—shock troops. Individual units, of course, became known for effectiveness after the fact. The cosmopolitan colonial manpower sources of Britain and France produced a wide variety of troops and of quality. Many units from India and Africa proved unsuited to European war. The most dramatic example, a new legend which revealed the linkage between a parent society and the nature of an army, was that of the Australians. The "diggers" became known as a virtual large-scale corps d'elite, to an even greater extent than the highly effective volunteer contingents from Canada and South Africa.

The Australian stereotype was born in 1915 on the beaches and in the hills of the Gallipoli peninsula. The Allies, under the urging of First Lord of the Admiralty Winston Churchill, were trying to use their naval advantage in flanking the Central Powers through the straits leading to the Black Sea. It was hoped that Turkey might be knocked out of the war and that an all-weather supply route would be opened to provide for the half-armed Russians. The campaign employed British, French, and Australian and New Zealand troops and saw some of the least imaginative generalship in a war distinguished by sub-mediocre performances. Nevertheless, the Australians in their turned-up bush hats, lanky, casual, fierce in the attack and resolute in the defense, caught the eyes of the world. The British and French were somewhat bitter at the limelight given to the diggers. The newcomers to the world scene, from an exotic stereotyped country and romantic origins, descended from convicts, were also thorns in the side of military orthodoxy. All volunteers, they symbolized individuality in a war

27

of mass armies. Their officers held rank by cooperation and common sense, and the troops were much given to thumping military police, rowdyism, and open contempt for "pom"—home British—officers, discipline, hierarchy, and methods. The diggers loomed larger than life in the Middle East and on the Western Front, maintaining their reputation in the Second World War, in Korea and Vietnam. In the final battles of 1918 in France and Flanders they were the "fire brigade" that halted the German spring drive. Comprising only nine percent of the British forces in the west, they took over twenty percent of German prisoners, guns and territory taken by all Allied forces in the offensives launched from April to October 1918,[12] and more than the Americans. In the Second World War the diggers' reputation for aggressiveness and casualness toward "pom" discipline and courtesy continued. At the siege of Tobruk the Australians created another legend; their role in New Guinea was obscured by the Americans and MacArthur, even though Australian casualties were double those of U.S. troops. The Japanese later saw the Australian militia at Milne Bay as the force that "knocked the first chip out of the Japanese war sword."[13]

Winston Churchill spent much time and effort wrangling with Australian Prime Minister Curtin trying to get Australians back into the Mediterranean theater after Pearl Harbor. But perhaps the highest tribute to the quality of Australian forces was the enthusiasm with which American troops praised Australian martial prowess, while being contemptuous of most other Allies. As brilliant as the Australian performance was, it also created problems for Australian national defense, a certain casualness and overconfidence. The American presence saved Australia from invasion in 1942, and the Australian defense posture in the Far East leaned heavily on the commitment of Britain and more recently the United States. Confidence in legendary courage on the part

28

of an elite is all too durable, and tends to break up only under the hammer blows of decisive defeat. The French *chevaliers* took a century to get the message from the English longbow. The urge on the part of politicians to portray war as a remote spectator sport is not an exclusively Australian affliction. But the defense and foreign policies of that remote outpost of the West have been long dazzled by their own mythology in a way that is becoming the pattern in other industrialized nations, as the burden of psychological and physical defense has fallen onto fewer and fewer shoulders. In that sense, the corps d'elite may resemble a high school football team which finds itself in a gang fight. In such a case, the spectator ethic produces disaster.

In a more general sense, however, more illusions were crushed than legends made during 1914-18. The First World War was an awful crucible for the West and spelled the beginning of imperial decline in much of the world. The crowns of Germany, Austria-Hungary and Turkey were hurled to the dust; the seeds of Communism, Fascism and Nazism planted. Faith in God and authority was shattered, and a perverted humanism arose, a faith in man alone in the universe, but man relying on action, not thought or reason. The images of those shock troops who had gone on in the face of the deluge of fire and iron that had wiped out a generation of Europe's youth came back to haunt the survivors and heirs. A great wound had been inflicted on the world, and the rise of the new corps d'elite was the scar tissue.

29

HITLER'S ONLY VICTORY:
COLLISION OF ELITES IN SPAIN

$\int$PAIN is a land that time and Europe forget much of the time, except that every once in a while the *furor Ibericus* blasts forth, and the world remembers the slumbering fires beyond the Pyrenees. Spain, nevertheless, is usually minimized or ignored in discussions of the flow of Western civilization. Its modern history is a blur to even its close neighbors—except for the years 1936-39, the Spanish Civil War. And even then the memories recalled are very selective. Yet that struggle did break through the barrier of apathy, and Franco, the victor, remains a stock villain in the repertory of Western liberals and Communists. The images of Spanish history lend themselves to easy stereotyping: the expulsion of the Moors, Ferdinand and Isabella, the Armada, Cortez and the *conquistadores,* the Inquisition. . . . But in the mind of the West, Spanish history trails off. Even the military zenith of the Spanish in Europe is forgotten, as are the campaigns of Napoleon in Spain and the birth of modern guerrilla warfare. The

30

Carlist Wars, the tensions between the Church and the land-owners and the liberal forces of Spain are forgotten, and the humiliation of Spain in the Spanish-American War is ignored even in America. Most Westerners think that Franco was a Fascist, when he was slightly liberal by Spanish norms.

The selective and uneven historical memory of ideologues of the Left, and, to a lesser extent, the Right, trapped many of them in the dry, barren valleys of Spain in the late 1930s. To them, forgetting a similar outbreak a century earlier, the Spanish Civil War was a thing unto itself, a morality play with live ammunition. And their weak memories cost them much when they forgot that they were really in someone else's quarrel, in a situation no one could understand clearly, and that both sides executed captured foreigners as a matter of course.

There has always been something about the air and the climate of Spain that does strange things to the perceptions of foreigners. It was no different in the Spanish Civil War, which became an ink blot onto which viewers projected their own needs and desires. As in the early nineteenth century, thousands flocked to the fray. Perhaps it was because the war was a goulash of ideologies, involving liberals, anarchists, Fascists, Communists, Nazis, clericalists and anticlericalists, separatists, royalists, and nihilists. There was something for everybody to identify with. The struggle had begun in the tangled overheating of the Spanish political system. The differences between political parties simmered and inflamed in the late 1920s and early '30s, like poisons in an infected appendix. The dictatorship of Primo de Rivera was succeeded by an elected government. But the tensions among all the groups needed to hold it together could not be resolved. The Inquisition was evidence of the readiness of Spaniards to judge each other readily and harshly. Finally, the Republic resorted to clandestine and gang-sterlike tactics to silence members of the Rightist opposition.

31

While the Rightists could not hope to gain an electoral majority, they knew that the liberal forces were faced with the same problem. And the Rightists had the Army, and, indeed, many of them were officers in it. Thus the Civil War began with a revolt of the Right against the Republic, and the Republican forces became known as Loyalists and the revolting group known as the Nationalists. The fierceness of the conflict saw the political differences explode into an orgy of bloody atrocities by all factions and bloodier battles, since zeal surpassed skill among most of the combatants. As with the American Civil War, it was a classic demonstration of the real danger in violent political rhetoric. In fact, so much attention was paid to the ideological intensity and the politics that many of the military lessons of the war were lost to the Western democracies in the next war. It has also been argued that the Germans learned the wrong lessons, that they built their air force toward the wrong goal, that of ground support. And others suggest that German aircraft designs were frozen too soon. Certainly the experience in Spain did the Italians little good. It may be, however, that Spanish "blooding" allowed the Germans to hone their blitzkrieg methods to a fine edge and thereby conquer far more of Europe in 1940-41 than would have otherwise been the case.

What was apparent in Spain was that decision in battle was determined by relative handfuls of combatants, the pilots, the artillery observers, the armored forces and infantry shock troops. This was a lesson already learned in the First World War and forgotten. But surprisingly little is retained, even by soldiers, of the lessons of conflicts. Key units bore the burden of battle in the early phases of the war because of the untrained condition of both sides. In the course of the war, as both sides tried to build fully trained armies, four elite corps entered the stage. The first was the Spanish Foreign Legion, an elite of nihilism, the "bridegrooms of death." The second were the Italian blackshirts, defined as an elite by Mussolini but by few others, and so irksome to Franco that he worked to have them removed soon after their arrival. The third were the nemesis of the Foreign Legion—the enthusiastic, idealistic International Brigades, whose spirit and courage made up for many of the defects, the backbiting and the stupidity of the Loyalist and Communist leadership. Last, and perhaps most decisive, were the men of the Condor Legion, the fighting portion of the German contribution, an elite force based on cool, professional response to a military problem, and most effective of all four because of technology, expertise, and lack of emotional involvement with the war itself. The German contribution to the Nationalists not only bought time but allowed the Nationalists an edge in training which eventually bought victory in spite of undistinguished tactics.

At the beginning, the Nationalists also had the upper hand with possession of the corps d'elite of the Spanish Army, the Spanish Foreign Legion, a trained, tough, fierce force, loyal to one of the leaders of the revolution—General Francisco Franco.

The *Legion de Extranjeros* had originally been formed as a response by relatively modern-minded, middle-rank Spanish officers to the deterioration of the army as a whole. Founded by Lieutenant Colonel Jose Millan Astray and the youngest major in the army, Francisco Franco, the Legion was organized in Spanish Morocco in September 1920: Millan Astray's first words to the volunteers—only a fifth of them actually foreigners—were: "You're here to die. Yes, to die."

Millan Astray and Franco met at a shooting match and discussed the need for a new kind of force in the Army. They then assaulted the creaking army hierarchy with demands for reform in infantry tactics to take advantage of the lessons learned by the combatants in World War I. Millan Astray developed the design

33

of a new unit while visiting the French Foreign Legion for three weeks. When they launched the Legion with official approval, liberal opponents feared it with some foresight, labeling it "the misbegotten child of an unbalanced mind."[1] After the opening of fifty-four recruiting offices, the Legion was established at a strength of three battalions *(bandera)*. Twenty of the first hundred volunteers were rejected on medical grounds. At the relatively well appointed Legion depot at Riffien, the piggeries supplemented the rations and made handsome profits for the Legion in the years to come. The Legion's standard of living, much higher than that of the Spanish Army as a whole, was at the level of other European armies of the period. Millan Astray and Franco devised a new tailored uniform with jodhpurs, open shirt collars, gauntlets, British-made web field gear, and white-lined capes. Any neatly maintained beard and hair style was permitted, and a special corps of camp followers was formed.[2]

The Legion saw its first action soon after the slaughter of 10,000 Spaniards at Anual. The Rifs were posing a serious challenge, and Spain fought alone against the rebels under Abd-el-Krim until joined by the French in 1922. The Legion became the work horse of the Moroccan campaigns, fighting in 845 separate engagements and expanding to eight battalions.[3] When Millan Astray was gravely wounded and his successor killed, Franco assumed command of the Legion at the age of twenty-eight. His reputation for coolness under fire, and his design of the amphibious landing at Alhucemas in the final victorious campaign in 1923, led to his promotion to brigadier general in 1926.[4]

In the 1930s, the Legion became a virtually independent force within the Spanish Army. It gained international notoriety when used with Moorish troops against the miners in the Asturias rebellion of 1932. The Legion and Franco, co-director of the operation,

A rare photograph of Francisco Franco, talking with a Kabila tribal chief in Morocco. He was then colonel in command of the Spanish Foreign Legion. A civil war would follow in Spain, and he would be-come head of state. (Photo courtesy of Spanish army)

were criticized when it was found that the rebels had unusually high casualties and that atrocities were committed after fighting ceased.[5] It is not surprising, then, that the Spanish Civil War began in earnest in 1936 with the Foreign Legion's crossing into Spain from Morocco by sea and air. The Legion served as a vanguard in the first phase of the Spanish Nationalist revolt—again with Moorish troops. In the 1937 Jarama battles, however, it collided bloodily with the International Brigades. During the war (1936-39) the Legion expanded to twenty battalions, including air units and Italian Blackshirts and French volunteers.[6] But, after 1937, the badly crippled Legion, its hard core decimated, faded from the spotlight.[7] The unit had truly lived up to its founder's battle-cry: *"Viva el Muerte!"* From 1920 to 1939 it lost forty-five percent of all its officers and thirty-eight percent of its enlisted men.

The Legion shared in the ceremonial glory of the Nationalist victory in 1939 and returned to Africa. (The pattern of blood sacrifice persisted; in 1956 a Legion unit made a last stand against Moroccan "irregulars," with the loss of almost a hundred men.) Franco, most durable of the strong men of the 1930s, became less and less identified with his days as the youthful firebrand, the Legion commander who had postponed his wedding to lead the Legion in battle.

Whatever his motives in co-founding it, Franco profited personally from the Legion in the enhancement of his career, the experience of actual combat and planning, and development of an image. Ultimately it became the cutting edge of the Nationalist revolt which made him chief of state. Franco has been characterized as cold and calculating, but stories—perhaps apocryphal—are told of his shedding tears when told of the Legion's ordeals and exploits. Of greater general significance, however, is the way in which the *Legion de Extranjeros* moved from fighting enemies foreign to enemies domestic, how personalized its allegiance ultimately was, and how its militant nihilism was exploited. Sharp tools must be used carefully.

Ironically, the Spanish Foreign Legion generated its own nemesis in the form of the International Brigades of the Loyalist side. The early Nationalist successes were a threat that untrained masses could not counter. The origin of the idea of International Brigades is somewhat murky. Hugh Thomas indicates that Tom Wintringham, British Communist military theorist, suggested their formation.[8] The five brigades ultimately formed under Comintern control (numbered XI and XV) consisted of 2,000 to 3,000 troops each and were made up of men from many countries. Beyond the thresholds of voluntarism and medical examination, there was no screening intensive enough to support the American Brigade veteran Alvah Bessie's view of them as "the best sons of the international working people of the world."[9] Nor was the performance of all the Brigades on the whole up to the standard of the *Legion de Extranjeros.* But the better units, the Germans, British and Americans, were. Many volunteers seemed to be in search of "self-discipline, patience and unselfishness—the opposite of a long middle-class training," i.e., existential self-definition,[10] not an uncommon trait of the members of the corps d'elite.

The Brigades were generally not well trained by European standards of the period, but a salting of veterans of the 1914-18 war and others often compensated for that deficiency. Weapons were not standardized. Medical support was inadequate. Estimates of overall strength range from 40,000 to 125,000, and percentages of actual Communists in the ranks also vary from forty to sixty.[11] Thomas thinks that eighty percent of the Brigades were working class, with sixty percent Communist before joining, and twenty percent remaining in the party afterward.[12]

37

The Nationalists' foreign support included 20,000 German advisers and Condor Legion, 40,000 Italians, and a smaller number of individual foreign volunteers. The best Brigade units—for all their deficiencies in training, leadership and inner political tension—provided their greatest service, especially in the fighting around Madrid in 1937 when they wore down the Spanish Foreign Legion. The success was at great cost to the Comintern, not only in terms of casualties but in the development of conditions which led to mutiny and desertion. Discipline varied among the national groups, creating major problems as the growing Soviet influence began to make itself felt in the Republic. The dark spirit of Stalinist purges and the secret police mentality caused reactions which finally led to a civil war within a civil war.

By late 1937 the activities of the NKVD against the Anarchists and other non-Stalinist Leftists had split the Loyalist ranks. Through intrigue and ineptitude, Communist leadership had caused the death in action of a third of the International Brigades,[13] which led to internal uprisings. The Stalinist response after the use of force, including tanks, was the creation of "re-education camps." The Brigades merged into the Republican Army in October 1937, filling the gap created by the defection of the Foreign Legion. Like the Red Army in 1942, the honeymoon ended with a return to older traditions of discipline, saluting, and rigid uniform standards. And the losses mounted. Three-quarters of the Brigades on the Ebro in 1938 were killed or wounded. There were prizes for those who went along. A military role in Spain proved most rewarding for European Communists and Leftists as well. Togliatti, Duclos, Gottwald, Ulbricht, Gero, Rajk, Tito, and Soviet World War II Marshals Malinovsky, Konev, and Rokossovsky—as well as twenty-four post-World War II Yugoslav generals, several Italian Socialist and Liberal leaders, and André Malraux—served in Spain.[14]

Colonel General Heinz Guderian, a key proponent of tank warfare and the originator of much of the tactical doctrine that governed the German army's devastating use of tanks in World War II. Under his direction, the first panzer divisions were organized and equipped, and staffed with a new elite Panzer Korps. He is seen here observing combat operations in Russia, 1944. The framework of bars in the photo was a common form of field radio antenna in the German army at that time, fitted to the Hanomag SdKfz 521 halftrack for use as a command vehicle. (United Press International Photo)

The rest is legend and tragedy. Many British and Americans who survived and escaped from Spain were denied a chance to train men or, in many cases, even serve in the bigger war against Nazism and Fascism. In the United States, veterans of the Brigades—the George Washington and Abraham Lincoln Battalions—while receiving legendary status on the Left, were on the U.S. Attorney General's subversive list until 1971. And they had much to pass on, for they had been through the laboratory of the blitzkrieg, punished by the ear-shattering sirens of the Stuka dive bombers, threatened by the treads of the Panzers and harried by the methodical use of various arms in a new kind of warfare that would soon shatter the armies of Poland, Denmark, Norway, France, Belgium, Holland, Britain, Yugoslavia and Greece. Their teachers had been the men of the Condor Legion whose power was built on speed and craft, not on mass and strength.

In the case of the Condor Legion, the purposes of German diplomacy in Spain were murky (even to the Nazis themselves) and the actual composition of their Condor Legion ever shifting. The men who served included volunteers for combat service and officers and technicians under orders. Overall estimates range from 20,000 to 80,000.[15] German teams of advisers—technical, security, and logistics personnel—outnumbered relatively small air and ground combat contingents. The term "Condor Legion" was often used to describe the entire German military support and advisory efforts; in reality, however, it actually referred to the fighting components. The combatants were relatively few at any one time, not numbering much more than a battalion on the ground[16] and about 200 aircraft of all types,[17] about 6,000 men in all.

In practice, the Condor Legion was more pragmatic than ideological. The wearing down of the Condor Legion in the 1938 battles led to a hard-headed exchange: Hitler agreed that the

Legion would be reinforced in exchange for a German percentage of interest in the Spanish mining industry.[18]

The impact of the Condor Legion was manifold. First, it helped the Nationalists attain a much higher standard of training than the Loyalists. It further acted as a visible contribution to the Nationalists in the face of Soviet support of the Republicans, and was far more useful than the more numerous Italian contingent. But it also provided grist for the propaganda mills of the Loyalists and their supporters in producing examples of German frightfulness, for example, at Madrid, Durango, Guernica and Almeria where the Germans were the instrument of mass bombardment of civilian populations on a new and frightening scale.[19] While the predominant sympathy of the world press was with the Loyalists, the Nazis were strangely oblivious to their vulnerability on this point, especially in their clinical—and cynical—attitude toward Spain as a laboratory, a view shared, but more subtly, by the Russians.

That the Condor Legion was more technical than ideological was reflected in the admiration of its members of their opponents' spirit which led to frequent observations toward the end of the war that: "We are fighting on the wrong side."[20] How much did the war in Spain serve as a training ground for the Germans? General Ritter von Thoma, one of the most experienced armor leaders in the twentieth century, served in Spain and wrote the primer for the blitzkrieg. Many others, including future General Staff principals Wilhelm Keitel and Walther Jodl, also served in Spain. The two top German fighter pilots of the early years of World War II, Adolf Galland and Werner Mölders, fought with the Condor Legion.

The war was, in terms of the investment, the Nazis' greatest victory—in the short run. The setback to Communism in Spain was severe, coming on the heels of the Great Terror and just before the Russo-Finnish War and the Nazi-Soviet pact. It shattered the underpinnings of Western Marxist intellectualism by showing the consequences of idealized violence and the unpredictability of history.[21] In that, Hitler destroyed better than he knew, but at a high cost in the long run. The momentary advantages of combat experience and economic gain were in the end offset, perhaps crucially. The image of the Luftwaffe as the killer of children was set. The freezing of German weapons systems design on the basis of Spanish combat experience, e.g., the Stuka, the Messerschmitt 109, and the light tank, may have delayed the development of the German jet fighter and left a gap in the Luftwaffe's strategy in air arsenal at a crucial moment in the Second World War. And when Hitler sought a serious commitment from Franco and a road through Spain to invade Gibraltar in 1942 to close the Mediterranean, he found two could drive a hard bargain.

In the end, Spain proved less disastrous for democracy outside Spain than many thought at the time. It was the source of many legends and artistic and literary images that marked the breakdown of the honeymoon of Western intellectuals and Soviet Communism. The cynical and heavy-handed oppressiveness of the Stalinists upset many. Russian weapons and men in Spain put an end to the credibility of Soviet pacifism. Few were willing to see the war as a purely Spanish phenomenon or as an affair, not of good versus evil but of men against men in excessive commitment to fallible causes. But the war did reverse the sense of inevitability in the growth of Marxism. In that sense the effect of the Condor Legion is best judged in the anguished post-mortems of Spain by Marxist military historians.[22] The Soviet Union became a nation, rather than the headquarters of an international movement. The Second World War saw further erosion of the Russian image as a bulwark of militant Marxism, shameless appeals to Russian nationhood, and a dissolution of the Comintern. But it was in the dry barren hills of Spain that Stalin, like Napoleon, met his first foreign setback, the first of a series, which destroyed the myth of Soviet Russia as the militant champion of the international working class. In the long run, it was Hitler's only victory.

IV

"MOBS FOR JOBS":
THE DECLINE OF
SOLDIERLY HONOR

THE traditional elites had stood in the forefront of battle. They were the first team of soldiery. The Gurkhas, the Guards, the Evzones, the Spartan 300, the Persian Immortals charged home in the face of certain death. They were rocklike in defense. But the gangster-warrior elites born of the Second World War ran about the rear areas of the enemy, destroying, confusing, and avoiding a fair fight whenever possible. "Mobs for jobs," the British called them. Private armies. Not soldiers, really. Nor civilians. Wore what they liked; developed their own vehicles and tactics. Drove higher commanders and the MPs wild. And their leaders usually had enough influence to keep them free of control by higher command. Like the airborne, they played against the second-string. Every act of harassment required a response by the enemy, tying down larger security forces, cutting communications. In that sense, they were guerrillas in uniform. The enemy was held in place in conventional warfare by the parent force and then jabbed and pinched until he bled from a thousand wounds.

In 1939, the German intelligence service had raised teams to fight behind enemy lines. They were known as the Brandenburgers, but their role was essentially passive. They often wore the uniforms of the other side as they disrupted, deceived, and gathered intelligence. The Allies moved in the direction of attacks and aggressive sabotage in the enemy's rear, beginning with the Commandos in 1940. Later, the Brandenburgers followed suit. The "mobs for jobs," while a symbol of individual aggressiveness, also compensated for defeat. Once the Allies regained the initiative, these special forces were either converted into conventional units or disbanded. No new models appeared after 1943. Only two species survived the war: the Commandos, and the British and Belgian Special Air Service. But they made their weight felt in 1941-42. The Germans had been forced to divert troops to cope with partisans in the Balkans and behind the lines in Russia, with Commandos and the Resistance in Western Europe. Eventually it tied up their strategic reserves, forty-five divisions in the Balkans alone by 1944. In the Far East, a whole garden of Allied "mobs for jobs" sprang up. In North Africa, British "mobs for jobs" tore up the rear of Rommel's Afrika Corps, wiping out hundreds of planes and vehicles and raising general hell. Hitler was impressed, just as the Allies had been by Crete. He ordered an expansion of the Brandenburgers and an increase in their operations.

And so hundreds of clean-cut middle-class boys on both sides were trained to slit throats, gouge eyes, break backs and crush testicles, showing remarkable enthusiasm for things that would have led to a death sentence and the revulsion of the community at home. It was not war. It was something worse. And the public on both sides lapped it up. Perhaps the special forces were a hint of the future, of a world of defunct morality and the triumph of brute power. They did much to undermine the idea that soldiers were agents of the state under close obligation to play by rules.

And the "mobs for jobs" did another thing. They put surprise back into business as a principle of war. The only problem was that sometimes they surprised the wrong side.

Creating forces for special operations was one way that higher commanders could keep close control of tactical units. Their very existence was a suggestion that the regular forces were inept and incapable of really bright, fast-moving operations. And they were usually seen as a nuisance and a drain by commanders ordered to provide them support and administration. Some, like the Australian Z Special Force and the British Force 136, 204 Mission, V-Force, and Operations Centre Khartoum, were far over into intelligence-sabotage rather than military operations. But most were closer to light or motorized infantry in organization and weapons. Just as storm battalion and tank enthusiasts saw armies as organisms with delicate nerves and circulatory systems behind a relatively thin skin, the advocates of "special units" viewed their creations as scalpels to be wielded against enemy lines of communication and critical points. But there was a difference. The storm troops and tanks opened up the enemy for conventional forces following. The "mobs for jobs" operated as harassers.

The first and most famous of this new breed were the Commandos. They were born of Churchill's combative enthusiasm in the dark days of 1940. And, as with the airborne, Churchill pushed hard to overcome War Office inertia and resistance from commanders of conventional army units. Churchill had called for a rapid-moving, aggressive corps of "storm troops" even before the 1940 campaign in France ended. Contemplating the arrival of the Australians in England—they were later diverted to the Middle East—he suggested a fire-brigade role with light vehicles and automatic weapons. They could respond rapidly to an invasion and eventually carry the fight to the Continent in the form of raids.[1]

This inspiration was soon reinforced by a suggestion from Lieutenant Colonel Dudley Clarke, a military assistant to the Chief of the Imperial General Staff, to form a volunteer raiding force. Churchill demanded 10,000 such "leopards,"[2] citing the example of German storm troops in 1918. The Commandos were the result. Their name was adapted from the Boer mounted bands in the South African War of Churchill's youth. In the summer of 1940, the period of gravest invasion fear, Churchill and the Commando unit leaders battled against opponents who feared loss of the army's best men and a decline of morale in regular forces if special attack units were formed. Commandos survived, with several purposes: to kindle the spirit of the attack, to gain battle experience, keep the Germans off balance on the Continent, put Britain on the offensive, and to serve as a laboratory for "triphibious tactics."[3]

Drawn initially from Regular Army units and then from Royal Marine volunteers, they were ultimately organized under the latter force. The first units formed around twelve companies of volunteers raised in early 1940 for the Norwegian campaign. These Commandos, who had intensive training in field craft, small-unit tactics, and raiding, wasted no time in getting action. They raided Boulogne in late June 1940 and Guernsey in mid-July 1940. The commander of Combined Operations, the headquarters that controlled the Commandos and their air and naval support, was Admiral Sir Roger Keyes, V.C. Meeting bureaucratic obstruction, Keyes concentrated on large-scale operations to make a large publicity splash. But the resulting inactivity caused morale problems, described in the novels of Robert Henriques and Evelyn Waugh, both Commando veterans. After the Lofoten and Spitzbergen operations in the fall of 1941 (the latter conducted by

47

Canadian troops, not Commandos), the charismatic and aristocratic Admiral (later Admiral of the Fleet Lord Louis) Mountbatten was appointed head of Combined Operations. His
youthfulness and drive were part of a British effort to substitute
grander raids for a full-scale invasion urged by the Americans (but
to be conducted by British forces) in 1942 to relieve the Russians,
in the war since June 1941. Under Mountbatten, the Commando
"empire" expanded rapidly. A large raid on the Lofoten Islands
off Norway came in late December. Operations were launched
against the French coast and in Crete and Syria, and North Africa
in 1942. Although the large and bloody "raid" against the Pas de
Calais at Dieppe in August 1942 was a Combined Operations
undertaking, few Commandos were involved. Eventually the
Commandos trained 25,000 men who saw action in Madagascar
(1942), Tunisia (1942-43), Burma (1945), and Germany (1945).
More and more, the Commandos were fielded as infantry in sustained combat, rather than as raiding forces. After the war, much
reduced Commando forces were continued in Britain, the Com-
monwealth, and in Belgium.

That the Commandos drained the British Army of leaders must
be weighed against their value as a symbol of Britain fighting
against evil triumphant in Europe. At the very least, problems of
discipline were reduced in the pre-invasion period by attracting
combative volunteers into remote areas for mountain and amphibious training, kept eager by the prospect of personal combat.
They did cause the Germans concern. Since the Commandos
generated no political virulence in following years, it is easy to
rate Colonel Clarke and Mr. Churchill as highly intuitive. But
there was a debit side to the ledger as well. Even the short, sharp
shocks of contact with the enemy exacted a price inordinate by
the casualty standards of World War II; 1,760 Commandos of the

25,000 trained were killed in action. And the Commando concept set a pattern for a new kind of war which regressed from standards of civility which had been evolving since the seventeenth century. From December 1941 to November 1942, the United States found itself in the same position as the British. Attacked and humiliated, they could not come to grips with the enemy on a large scale. The Commando model, a symbol of aggressiveness, was naturally attractive. In June 1942, the first Ranger companies, based on the Commando idea, were raised in Northern Ireland from among U.S. troops stationed there. The concept was approved by General George C. Marshall, U.S. Army Chief of Staff, at the suggestion of Brigadier General Lucian Truscott, head of the U.S. mission to Combine Operations in 1942. (Whatever the origin of the "Rangers" title—possibly Rogers's Rangers as depicted in the film version of Kenneth Roberts's *Northwest Passage*—no one seems to have noticed that Rogers fought for the British in the Revolution.)

The Rangers trained initially with the Commandos at Achnacurry, and their baptism of fire came in August 1942 in the large-scale raid at Dieppe. The Canadians provided the bulk of the force and suffered the most casualties. The Rangers returned unscathed, but their presence led American newspapers to suggest a stronger American contingent at Dieppe than was the case. That lucky beginning was not an appropriate augury, for the Rangers were to pay a butcher's bill for elite status.

The Rangers next saw duty in North Africa as a full battalion commanded by Lieutenant Colonel William O. Darby, who, like Frederick of the 1st Special Service Force, was viewed as a promising young officer. Darby's Rangers grew to the size of a light brigade by early 1943. Two battalions went ashore at Gela on Sicily in the first phase of the Sicilian invasion in July. They served

Major General Lucian Truscott, commander of the American 3rd Infantry Division and founder of the U.S. Army Ranger program. (U.S. Army Photograph)

as a special task force under Darby's command in Patton's drive on Palermo and captured up to 4,000 prisoners in one day. Still their luck held.[4]

In the invasion of mainland Italy in the late summer, the Rangers landed near Salerno and were involved in the drive on Naples. Although they had been selected, trained, and lightly equipped for hit-and-run roles, the Rangers were used more and more as conventional infantry. At Anzio, the odds of such misemployment caught up with Darby's force. During that landing, three Ranger battalions took Anzio with relative ease, the 1st Special Service Force on their right flank. They were then ordered to advance out of the beachhead to the east by night to spearhead the 3rd U.S. Infantry Division's attack on Cisterna. The 1st and 3rd proceeded slowly along a flooded irrigation ditch until dawn on January 30, 1944. One-half mile from Cisterna they were caught in the most dramatic ambush suffered by a major American force in the twentieth century. The Germans brought tanks

50

and self-propelled guns to bear at close quarters, raking the pinned-down Rangers at point-blank range. By noon, half of the Rangers were dead or seriously wounded. English-speaking German officers carrying hand loudspeakers used prisoner hostages to force others to surrender. Although ten percent of the Rangers captured escaped later, the two battalions were wiped out. Colonel Darby attempted to reach the beleaguered force with the 4th Battalion, which, after suffering fifty percent casualties and the death of all its company commanders, fell back. Only six returned of the 767 Rangers who moved against Cisterna on the night of January 29.

Darby had earlier refused a promotion to regimental commander in the 45th Division to stay with the Rangers, but after Cisterna he was without a command. Of the original 1,500, only 449 Rangers remained. Two hundred and fifty were transferred to the 1st Special Service Force and the rest sent home to train troops. Darby, promoted to brigadier general, was killed in action in early 1945 while executive officer of the 10th Mountain Division.

The remaining Ranger units—six battalions in all were formed—met further frustration and some success. The 6th Ranger Battalion participated in the spectacular raid which rescued American prisoners of war at Cabanatuan in the Philippines. But fate played another card against the 2nd and the 5th Battalions at Normandy on D-Day. Two companies of the 2nd Rangers lost 62 of 130 men crossing "Dog Green" on Omaha Beach. The 5th Battalion lost 135 of 225 scaling the hundred-foot sheer cliffs at Pointe du Hoc, only to find that their objective, the German heavy artillery emplacement, had had their guns removed.[5]

After the Second World War, the Ranger concept lived on in the U.S. Army, but without a permanent organizational base. Some American divisions formed their own "Ranger" units, and battalions were used in Korea. In the mid-1950s, however, Ranger units were disbanded. Special training was conducted for individual volunteers, leading to the award of Ranger insignia. In 1968 the U.S. Army revived the Ranger unit concept again in the form of a special provisional battalion. The image persisted, though the promise had miscarried.

Another offshoot of the Commando concept was the U.S.-Canadian 1st Special Service Force, a product of Admiral Mountbatten's enthusiasm. The ultimate among corps d'elite in terms of confusion and tangled military bureaucracy was created for a mission which quickly evaporated. Like Merrill's Marauders, the 1st[6] was a product of the Quebec Conference and is a bizarre footnote in the history of Allied military policy. With the support of Admiral Mountbatten—then chief of Combined Operations—a scientist, Geoffrey Pyke (who also suggested huge aircraft carriers made of "Pykrete," a mixture of sawdust and ice), proposed using highly mobile tracked vehicles in a massive raid to destroy Norwegian industry.

As a result, the United States and Canada each raised complements and merged them in the 1st Special Service Force at Fort William Henry Harrison near Helena, Montana, under the command of a young American officer marked as a "comer," Lieutenant Colonel Robert Frederick. While the Canadian half of the 1st was hand-picked, the American half was a collection of marginal types culled from stockades and unit rejects. Many of those initially assigned were low on the scale of intelligence standards in the U.S. Army. Extensive reselection was required. Frederick made regular trips to Washington to fight bureaucratic battles, and steadily formed the unit into his own creation, adopting red berets and crossed arrows as distinctive insignia.

After receiving airborne, mountain and arctic training (the unit never jumped in combat or operated in high mountain or snow country), the 1st was threatened with disbandment. The Norwegian government in exile vetoed Pyke's scheme, frowning on wholesale destruction of its economy. But Frederick succeeded in putting off the dissolution. The 1st was then committed—without its "Weasel" vehicles—to operations in the Aleutians, in the Winter Line in Italy, at Anzio, Rome, and finally in the Rhone Valley. As the result of casualties and the disappearance of need for its special skills, the 1st was disbanded in December 1944. Most of its remaining veterans were assigned to airborne units.

The 1st Special Service Force was later claimed as a success in terms of U.S.-Canadian relations. Due to the international nature of the 1st, considerations of its mission, size, and employment took up an "inordinate amount of high level consideration."[7] Resolving the relationship of Canadians to U.S. command authority generated extensive legislation in the Canadian parliament. The senior Canadian officer in the 1st was authorized direct communication with Ottawa and the Combined Chiefs of Staff. U.S. and Canadian diplomats also spent much time discussing the employment of the unit. To further this hothouse effect, the 1st was administratively isolated from British Army channels in Europe, which led to their petitioning specially for British decorations.[8]

The symbolic value of the "Devil's Brigade" and the charismatic skill of Colonel Frederick kept the unit alive beyond its utility. Like the Lafayette Escadrille and the Eagle Squadron, the Devil's Brigade was a demonstration of international cooperation at the unit level. One of the great ironies of the twentieth century has been that even the peace-loving are enthusiastic about military legitimization; that is, the granting of respect and status through armed service. In such cases as the Fighting 69th of New York, made up mainly of Irish Americans, the French Foreign Legion,

53

At far right, Brigadier General Robert T. Frederick, commander of the 1st Special Service Force, a unit that combined American and Canadian troops and whose combat success led to the development of the American Ranger and Special Forces programs. Frederick is shown here at a briefing outside Rome in June 1944. The lieutenant general to his right is Mark Clark, commander of the Allied Fifth Army. (U.S. Army Photograph)

the 442nd Regimental Combat Team raised from Nisei volunteers, and the Anzacs, respect was gained by the success of arms. Early in Vietnam, American liberals who ordinarily detest war and all things military cited the achievements of black troops as a rationale for acceptance of American Negroes into the parent society. Using proven skill at homicide and destruction as the index of honor may seem a paradox. Such a rationale accepts war as part of civilization, rather than the antithesis of it.

Such "private armies" as Frederick's flourished east of Suez as well. Tropical warfare has generated a rich variety of special forces. Someone will no doubt notice a similarity to the pattern of generalist sparrows in the temperate zones and quetzal specialists in the most stable equatorial regions. The flowering of these strange species may be due to the greater effort needed to maintain operations against the barriers of vegetation—or lack of it in deserts—mountains, climate, and so forth. Troops operating in these regions in the world wars were often locally raised and far from home headquarters. The First World War saw Lawrence of Arabia and Paul von Lettow-Vorbeck in East Africa drive their opponents wild, even though they were fighting at great odds, with anything but elite troops. In the Second World War, however, corps d'elite abounded in Africa and the Far East, and the contenders for Lawrence's title were not lacking. Lowell Thomas, who made Lawrence's reputation with a film-lecture series, went to Burma to keep a close eye on Wingate, who came the closest. But there were others, some straightforward and functional, others more obviously a grandstand play for glamour and freedom from military routine.

The first of the tropical species to appear on the Allied side[9] was the Long Range Desert Group (LRDG). In 1940, British Empire forces in North Africa numbered less than 20,000 combat effectives against an Italian force of 300,000. Dry rot in Fascist militarism had been visible in Ethiopia and Spain and in the attempted invasion of France in June 1940, where six French mountain divisions hurled back twenty-four Italian divisions. But the odds were still impressively one-sided. British air power was also stretched thin, with home priorities sapping the quality of overseas units. Lack of effective air reconnaissance led Brigadier Ralph Bagnold to approach General Archibald P. Wavell, then commander of British Middle East Forces, with the idea of forming a special ground reconnaissance force under his command. Bagnold, a career Royal Engineer officer and brother of Enid Bagnold, author of *National Velvet,* did field research in the North African deserts in the 1920s and 1930s and was a noted author and academic expert on sand movement and terrain. Wavell gave his approval, and thirty New Zealanders formed the core of a unit that eventually grew to three hundred, supplemented by volunteers from Rhodesian and British units, as well as twelve Americans. The LRDG carried out over two hundred operations, including transporting spies, inventorying enemy movements and supplies, and liberating POWs. As the RAF gained air supremacy in North Africa, the LRDG shifted more and more to raiding and sabotage. LRDG members later participated in special-force work in Italy, Greece and the Greek Islands, Albania, and Yugoslavia. The unit, disbanded in 1945, was commanded by Colonel Guy Prendergast during 1941-43.

The LRDG was only one group which captured public imagination in Britain and the U.S. in the desert war. The Special Air Service Brigade and Popski's Private Army also operated independently and with the LRDG. The SAS was formed by a young major, David Stirling of the Scots Guards and a Commando as well.[10] The name of the unit was a cover for its basic purpose, deep-penetration raiding and sabotage, and for its size, which was about a platoon initially. Popski (an alias of Vladimir Peniakoff)

Lieutenant Colonel Vladimir Peniakoff, "Popski" of "Popski's Private Army," in Tunisia in 1943. (Photo courtesy of Imperial War Museum)

aimed at a role more akin to the German Brandenburgers, one of deception and sowing confusion.[11] The SAS suffered heavy casualties and disappointments when several major schemes miscarried, most notably at Tobruk. Stirling was captured in 1943 in Tunisia. As with other light elite forces, its shift to serious prolonged fighting was costly. The smaller raids of the SAS were proportionally far more effective.

These groups had a flavor of buccaneering about them to friend and foe alike. Their multinational, multilingual personnel, their contempt for even the casual uniform regulations of the 8th Army, and their customized vehicles and esoteric skills were colorful offsets to the weary series of British setbacks and disasters from April 1941, when Rommel arrived in North Africa, to the great victory of the 8th Army at El Alamein in November 1942. As in other theaters, when the major forces were doing badly, the firefly "mobs for jobs" lit up an otherwise very dark night. In spite of the SAS's bad luck, they proved their worth by destroying hun-

Jeeps of the Special Air Service, on patrol in the Western Desert, 1943. Special Air Service has frequently been confused with another British unit, the Long Range Desert Group. While the latter performed skillful reconnaissance and communications missions, SAS units conducted most combat operations behind enemy lines. The farthest vehicle in the photo has 30-caliber twin Vickers K guns mounted aft, and a 50-caliber Browning over the bonnet. There were usually Thompsons and Brens aboard as well. SAS dealt out an awesome amount of destruction. (Photo courtesy of Imperial War Museum)

dreds of German and Italian aircraft. The name was later applied to an airborne brigade designed for behind-the-lines work in Europe, composed of one Belgian, two British, and two French battalions. They remained a unit in the British and Belgian armies after the war, and their green beret was eventually adopted by the U.S. Special Forces.

Farther east of Suez, there was less publicity and more miserable fighting conditions, and a fiercer enemy. The price of glamour was high. The most visible elite forces in the Pacific area early in the war were the United States Marine Corps Raider battalions. The marines of other countries are relatively small and highly specialized, e.g., the British Royal Marine Commandos and the recently formed Soviet "white berets." But the U.S. Marines became a major field army in World War II. The Marines built strong links in Congress early in the nineteenth century and fended off repeated assaults by other services and the executive branch to pare down or eliminate the Corps.[12] Whether the Marines in early World War II attracted volunteers who should have served in more appropriate roles as leaders in other services, and whether U.S. Marine training produced higher quality troops, will long be argued. And, undoubtedly, the fact that the Marines were the first U.S. ground forces on the offensive in the Second World War attracted fighters. As the 1st Marine Division stormed ashore at Guadalcanal in August 1942, Marine pugnacity was further demonstrated in a raid against the Japanese on Makin Island. The raiding force was one that soon captured the American public's attention, the Marine Raiders[13]—so fearsome that legend has it that Mrs. Roosevelt, the President's wife, suggested prolonged quarantine before returning such men to civilian society. The Makin raid on August 27-28 was soon dramatized and magnified in the film *Gung Ho!* Two companies of "Carlson's Raiders," a little over two hundred men of the 2nd Raider Battalion, were

59

transported to the Japanese-held atoll in two old oversize submarines. The actual impact on the Japanese was slight, and losses of men and equipment were higher than expected. It was, however, forward motion after many defeats and retreats. Like the raid of B-25s on Tokyo led by Jimmy Doolittle, it served as a propaganda tonic. And one of President Roosevelt's sons, a Marine reservist, was executive officer of the raiding force commanded by Lieutenant Colonel Evans Carlson.

The Raiders had been formed at the urging of the President (and his son James) as a counterpart to the British Commandos. Marine regular officers resisted the idea of an elite within an elite, and argued that they were already in the raiding business. But the

Lieutenant Colonel Evans Carlson, left, commander of the Marine Raider Battalion, displays a flag seized during the spectacular Makin Island raid in September 1943. At right is his second in command, Major James Roosevelt. (Defense Department Photo)

units were formed nevertheless, and soon proved more glamourous and popular than expected—but not always as raiders.

The 1st Raider Battalion, headed by Lieutenant Colonel Merritt Edson, raided Tulagi and Savo Island in August-September 1942. Then, in the classic pattern of misuse, they were committed to sustained combat on Guadalcanal to the point of shuffling, mumbling exhaustion.[14] On September 8 the 1st raided from the left flank of the U.S. position, destroying Japanese supply dumps and defecating on Japanese food stores.[15] Soon afterward four hundred Raiders of the 1st beat off a major attack by 1,800 Japanese. Then, in slow, relentless hunting expeditions, Carlson's Raiders left the Guadalcanal perimeter on November 12, 1942, and returned on December 4. In a 150-mile march they pursued the Japanese 228th Regiment, fighting a dozen actions and killing approximately five hundred Japanese at the cost of sixteen Raiders killed and eighteen wounded. It was not war; it was a hunting expedition.

The Raiders remained organized until January 1944. Gallons of ink were spilled in the form of official memos arguing their organization, equipment and overall purpose. The Army and the Navy each made feeble attempts to gain control of the program. Four battalions in all were formed, plus a Raider regiment, which frequently cooperated in action with the Paramarines. In June 1943, the 1st Raider Regiment and the 1st Raider Battalion cooperated with Army troops on New Georgia and were later joined by the 4th Battalion. Casualties by August had mounted to half the units' original strength. Neither unit saw action again.

The 2nd and 3rd Battalions went into action on Bougainville in the northern Solomons in September 1943, along with the 3rd Marine Division. By this point in the war, their role was so interchangeable with regular Marine units that the 1st, 2nd, 3rd, and 4th Raider Battalions were incorporated into a Marine regiment to replace the unit lost in the Philippines; the 5th and 6th Battalions were scratched from operational planning. The Raider Battalion Training Center at Camp Pendleton was closed down. As with the "mobs for jobs" in Europe, the momentum of victory favored the big battalions.

The Raiders were an elite within an elite, volunteers stiffly trained, commanders hand-picked and given a free rein to develop units.[16] Like the Marine Parachute Regiment, the Raiders' exact casualty figures were not published; like some airborne units in Europe, they were used in prolonged combat, although trained and equipped for short, intense actions; and like other "private armies," they were not popular with other units.[17] Colonel Edson of the 1st went on to win the Medal of Honor and retired as a major general. The more publicized and controversial Carlson[18] had laced his tactical ideas with ideology absorbed while he observed with the Communist 8th Route Army in China in the late 1930s. Carlson was wounded on Saipan in June 1944, a year after the Raider battalions, including the more recently formed 3rd, were formed into the old 4th Marine Regiment, lost in the Philippines. He became a brigadier general before his death from a heart attack in 1947.

Problems of misuse, confusion on the part of commanders as to function, heavy casualties, and inter-unit tensions appeared. Other Marines, sensitive of regimental pride, saw the Raiders as publicity seekers, for the Raiders' greatest successes were gained out of sight and behind enemy lines. As with most "private armies," erosion of need, small size, misuse, casualties and animosity led to organizational extinction.

The Marine experiences, ironically, had little impact on the U.S. Army. Just as U.S. Rangers learned amphibious warfare and raiding from British tutors, the army began long-range jungle raiding not down paths blazed by Edson or Carlson but along a trail hacked by the British innovator, Orde Wingate, and his "Chindits."

Chindit is a corruption of "Chinthe," the half-lion, half-eagle Burmese temple gargoyle which Wingate chose as a symbol of air-ground cooperation. The British Long Range Penetration Force was raised in 1943 from British volunteers in India. Wingate, with exotic exploits in Palestine and Ethiopia on his record, gained Churchill's ear through Liddell Hart.[19] He insisted that only British and Gurkha troops and Royal Air Force ground teams be in his force. His anxiety over using Indian forces in a deep thrust into Japanese territory was not paranoid. Some Indian units had performed poorly in Malaya, and the Japanese had raised an "Indian National Army" from prisoners of war. More practically speaking, the Chindits would have to master English-language training quickly and to fend for themselves outside the British lines.

The results were uneven. In their first major operation, a long march into north-central Burma in 1943 supplied wholly from the air, eighteen percent of the 3,300 Chindits became combat casualties and twelve percent were laid low by disease. The refurbished and retrained Chindit force—increased to six brigades and officially designated "Special Force"—underwent another ordeal in the spring of 1944. After landing in gliders behind Japanese lines in north Burma and linking up with elements that had marched 400 miles, Wingate's command was thrown into confusion by his death in an air accident on March 24, 1944.

His successor had no clear instructions as to his mission. He did not know whether the Chindits were to interdict Japanese communications supporting attacks on Kohima and Imphal, to generally harass, or to spearhead an advance to the Irrawaddy. South East Asia Command put the Special Force under American

Brigadier General Orde Wingate (in sun helmet) briefs Chindits in Assam. The Chindits were Chinese troops trained by Americans. Together with Merrill's Marauders and other elite units, they conducted guerrilla operations against the Japanese in Burma and Southeast Asia. (Photo courtesy of Imperial War Museum)

General Joseph Stilwell's control on May 16. Wingate insisted that his men be kept in action only ninety days, but they now joined "Merrill's Marauders," an American long-range penetration unit originally designed to fight at their side. Stilwell's plan to capture Myitkyina in north Burma with Chinese troops had failed. The attack bogged down in the face of weaker Japanese forces and became what Field Marshal Slim called an "untidy, uninspired, ill-directed siege."[20]

Although he had declined earlier British offers of help, demands by Stilwell for British troops produced a collective disaster for the Special Force and Merrill's Marauders when "Vinegar Joe" kept them in action far beyond the limits of effectiveness. Only a direct warning from Admiral Mountbatten, head of South East Asia Command, underlining Stilwell's personal responsibility for keeping helpless men in combat, jarred Stilwell back to sanity. The damage, however, had been done. Most of Special Force and Galahad Force—Merrill's Marauders—were reduced to human scrap.[21]

Galahad Force was a child of the Quebec Conference. Raised in late 1943, it represented the U.S. Army's venture into long-range penetration. Since the British had an edge on experience, and since Empire troops comprised the bulk of the forces in Burma, General Marshall ordered the creation of Galahad, known first as the 1688th Casual Detachment and then as the 5307th Composite Unit (Provisional). The 5307th (survivors were phased into the 475th Infantry Regiment in the fall of 1944) was designed in view of British experience as a "throwaway." Volunteers for Galahad were later resentful, since the ninety-day planning guideline had become a promise. A joking reference during training by the commander, Brigadier General Frank Merrill, to a post-operational party in India did not help, all of which led to disaffection when the 5307th was kept in combat considerably

65

beyond the three months rule-of-thumb limit.[22] They never knew that they had been expendable from the start.

A call for volunteers produced a broad range of personnel. Jungle-experienced troops in MacArthur's command did not respond, so that most of the quota from the Southwest Pacific area was filled by men without battle experience.[23] A veteran said of the rest that the 5307th "contained far more than its share of drunkards, derelicts, guardhouse graduates, men of low intelligence and out-and-out bad men of the traditional kind."[24] The 5307th was raised primarily from volunteers among U.S. forces stationed in tropical areas, e.g., Panama and the West Indies. It was organized as a regimental combat team in three battalions, plus support—about three thousand men. Galahad trained under control of South East Asia Command in Burma in the late fall of 1943, since it was slated to cooperate with British Empire long-range penetration forces there.

Originally, Marshall intended the force to be trained and fielded by Wingate. But tense and tangled command politics in mainland Asia thwarted his design.

In the long run, Galahad's elite image was inflated by journalism, military folklore, and their identity as the only American ground troops in sustained combat in Burma. The unit was dubbed "Merrill's Marauders" by *Time* correspondent James Shepley when Brigadier General Frank Merrill, a Stilwell protégé, took command just before commitment to action. The undisciplined behavior of its members, like that of other corps d'elite, seems contradictory. Are not elites the best? But what is best in military terms? True, cutting throats and furtive vandalism hardly mesh with the laws of war. And there is also the volunteer effect. The call for volunteers within armies often produces negative selection. Unit commanders may respond to forced levees and mandatory volunteer quotas by sending their least desirable men.

66

This effect, for example, forced restaffing of the American contingent of the 1st Special Service Brigade. Beyond all this, it is naïve to expect aggressive fighters to be docile conformists.

John Masters, a senior officer in Special Force, painted the most vivid picture in *The Road Past Mandalay* of the results of the extended campaign in Burma among British troops. In the case of the Americans, conditions were similar, and perhaps worse in the long run. For Galahad went into the field without an organized medical evacuation system. The men became so resigned to dysentery that one company cut the seats out of their pants.[25] After several actions in north Burma, Stilwell committed Galahad to a major push to break the stalemate at Myitkyina. But the Japanese held on with grim steadiness. By late May the Marauders were in a pathetic condition. A battalion commander fainted three times in one day. Frank Merrill—exhausted from fighting the Japanese, the command, morale problems, and the climate—suffered two heart attacks and was evacuated. A malaria outbreak reflected, an inspector-general report noted, an "almost complete breakdown in morale," for simple individual precaution could have prevented it. By late summer the area around Myitkyina had become Verdun-like. The devastated countryside, the ghastly appearance of exhausted, hollow-eyed, sunken-cheeked survivors living on vitamin-poor emergency rations, the intense fighting, and the stench of death were horrendous. Replacements were hurriedly assembled in the United States and flown to Burma. The hellish scene made half of them major psychiatric casualties the day they arrived.[26]

Stilwell was a stern taskmaster. Before Myitkyina, Galahad marched five hundred miles. By late April almost the entire force had dysentery or malaria or both. Regular medical evacuation rules were waived as Japanese resistance stiffened. As a further measure to compensate for the failure of his Chinese troops, he ordered minimum evacuation and a dredging of hospitals for marginal effectives. Fifty of the two hundred Marauders ordered back to duty were immediately returned to the hospital by officers of the 5307th.[27]

But all ordeals end, and the Marauder remnants, less than a third of the original force, were pulled out of the line. The situation in the rest camp led to virtual mutiny. Stilwell visited the unit, just prior to his return to the U.S., to talk to spokesmen of the aggrieved Marauders, and entered their camp, a place that MPs refused to patrol. It was a tribute to his courage that, alone, he faced men who had been subsisting on local highproof and marijuana-laced "bull-fight brandy."[28]

The casualty rate in Merrill's Marauders was five percent less than had been projected, evidence of War Department planners' foresight, if not wisdom. The long-range impact of these disasters was reflected when no counterpart to Chindits or Merrill's Marauders was established in the postwar British and U.S. armies.

Most of these "mobs for jobs" were ground forces. But air war was a more fertile ground for the creation of corps d'elite. Pilots were an elite caste. The air forces, from the beginning, were known for their chivalric values, especially among fighter units. Individual leadership, style, sporting behavior, and freedom from the bureaucratic control of headquarters saw the emergence of such groups as Richthofen's Flying Circus. Airwar, 1914-18, seemed like a return to the days of knightly single combat. The aircraft sported gaudy heraldry, and, like the *chevaliers,* many labored to support a few highly trained young warriors. The refinements of the metal trades in the fourteenth and the twentieth centuries focused on making vehicles of war for a select caste bent on dueling to the death. When the men of the Eighth Air Force donned body armor in 1943, the wheel turned its full cycle.

There were many elite air units: for example, the RAF Path

Finder Force; 617 Squadron (the RAF's dam busters); U.S., British and Japanese carrier pilots; the *Stalinfalken;* the Luftwaffe's Squadron of Experts (JV44), the first jet fighters; the "Abbeville Boys" (JG26); and *Nachtjäger Geschwader* 4, a night fighter unit which shot down 579 Allied bombers, losing only 450 air and ground casualties during the war. The most renowned American Negro unit in the Second World War was the 99th Fighter Squadron.

The American Volunteer Group, however, a fighter group popularly known as the "Flying Tigers," the best known air unit to date, was an elite drawn from an elite, screened further by training and combat. The AVG's experiences paralleled those of small ground elite forces. Courage and skill filled in for deficient equipment; jealousy abounded, and in the end the unit was disbanded after many casualties and much disaffection.

The AVG had a strong visual image in its main operational aircraft, the Curtiss P-40B, which had a large ventral radiator scoop painted with a tiger's snarling mouth, ostensibly to frighten Japanese pilots. It made dramatic photographic copy in contemporary periodicals. A second-line fighter plane by Allied standards also characterized Western aid to China: symbolic, obsolescent and uneven. Several nations were involved in aiding China in the late 1930s when Chinese Nationalists and Communists faced each other over gunsights while keeping the Japanese at bay. The British, Americans, Germans, Russians, and Italians were all involved at one time or another in helping the Nationalists, whose units kept the bulk of the Japanese Army occupied from 1937 to 1945.

In the mid-1930s the Italians had trained the Nationalists. They awarded pilot status to all volunteers, thus pleasing powerful members of the Nationalist government who pressured Chiang Kai-shek when their sons failed flight training. The results were

69

grotesque and tragic, with bloody crashes in training and at demonstrations. Few survived in combat with the Japanese, who, as a result of easy victories, soon grew casually confident. The Russians replaced the Italians in 1937 and provided effective support, but their withdrawal in late 1939 left the Chinese without air cover or defense. It was not until late 1940 that the United States and Britain moved to support the Chinese Nationalist forces.

A Chinese mission to the United States succeeded in gaining American money, equipment, and manpower to form an illegal— by U.S. law—volunteer force, the American Volunteer Group. Recruiters were allowed to visit U.S. Army and Navy bases to recruit pilots, offering a lucrative one-year arrival-to-departure contract. Sixty-three Navy and thirty-eight Army pilots signed up, sailing for China on a Dutch ship with U.S. naval escort on June 9, 1941. Their commander was Claire L. Chennault, a former air force officer and a veteran—and a victim—of intraservice bomber-fighter faction fights. He had been in China as a mercenary pilot during 1937-39 and was a close friend of the Chiangs.[29]

The AVG had already been carefully screened on medical grounds as well as by training and performance. In China they were screened further, and the eighty-four left were then trained secretly at British bases in Burma by Chennault in his tactics, which emphasized the strengths of the Curtiss P-40B against lighter, more maneuverable Japanese fighters.[30] The AVG went into combat in Burma in December 1941, two weeks after America entered the war. In the seven months of its existence, the AVG destroyed 299 Japanese planes confirmed and 253 probables, and lost 12 planes in the air and 61 on the ground. Of twenty-six AVG pilots lost, ten died in accidents.[31]

The gap between militarism and military effectiveness was dramatized when the AVG was absorbed into the U.S. Army Air

Major General Claire Chennault, center, commander of the U.S. 14th Air Force in China and founder and leader of the American Volunteer Group known as the Flying Tigers. (U.S. Air Force Photo)

Force. AVG pilots, not under military discipline, required deft handling. Chennault tolerated much indiscipline, including one virtual mutiny.[32] Due to the overbearing manner of a visiting Air Corps general and USAAF physical standards imposed on fatigued men, only five of the AVG veterans received Army Air Force commissions. Others went back to the Navy and Marines or took lucrative jobs with Chinese and Indian airlines.[33] Chennault went on to command the U.S. 14th Air Force in China, whose men wore a Flying Tiger patch, and became involved in the tangled political infighting that swirled around the Stilwell-Chiang relationship. Chennault proved with the AVG that command experience, insight into personnel, relatively high pay, and careful selection could produce an outstanding unit even with

Some of the Tigers that flew. Here are Curtiss P-40s of the 16th Fighter Squadron, 51st Fighter Group, at a base in China in 1942. (U.S. Air Force Photo)

inferior equipment and without conventional military discipline. After 1945 Chennault carried the image with him when he formed a commercial air freight service, the Flying Tiger Line.

In spite of the setbacks and frustration, these wild varieties of corps d'elite did gain many successes and were powerful symbols as well. The Japanese response was sporadic, and the rigidity of their command structure suppressed such deviance. Their loss of initiative eventually generated another wilder kind of response. But in the European theater, Hitler attempted to fight fire with fire, and did very well indeed. As his point of departure, he used the so-called Brandenburgers, formed in October 1939 by the chief of German intelligence, Admiral Canaris. Their mission was intelligence activities, i.e., clandestine raids and infiltration. Like the British Long Range Desert Group and Special Air Service, the

Brandenburgers operated in small groups detailed from the parent organization.[34] Less visible than their Allied counterparts, they first saw action in the campaigns of the spring of 1940. Later they served in Africa, on the Eastern Front, and in the Balkans. A Brandenburger team was even landed in Southwest Africa from a submarine. They included recruits from non-Aryan ethnic groups, e.g., Indian army deserters and POWs, Afghans, and Balkan Muslims. In late 1944 the organization, which had grown from company to brigade size, was merged into the SS Division *Grossdeutschland.*[35]

The Brandenburgers became visible when they assumed a more aggressive role operating in small detachments under command of "super hero" Otto Skorzeny, an SS junior officer and an Austrian who caught the eye of Hitler as the Führer ordered the expansion of German special forces in 1943, in response to British operations in Europe and North Africa.[36] Skorzeny, a maintenance officer in SS Division *Das Reich* and in the *Leibstandarte,* directed several hairbreadth operations. Tall, muscular, and saber-scarred, his escapades matched his formidable appearance. Skorzeny's first triumph was the September 1943 rescue of Mussolini from a remote mountain resort at Gran Sasso in Italy closely guarded by pro-Allied Italian forces. Skorzeny, with a team of paratroops and Brandenburgers, arrived by glider, rushed the astonished guards, and achieved a relatively bloodless coup. Then, he and a pilot flew Mussolini out in a light liaison plane, barely clearing the impinging peaks. This affair made Skorzeny a matinee idol among the Nazi hierarchy.

In his next operation, with Brandenburgers and paratroops, he kidnapped Admiral Horthy in Budapest to stop the Hungarian government from seeking a separate peace. As impressive as these coups were, it is ironic that Skorzeny's greatest tactical failure, during the Battle of the Bulge in December 1944, had the most

73

strategic success. Promoted to *Sturmbannführer* and personally decorated by Hitler, Skorzeny received command of the 150th Panzer Brigade in late 1944. Brandenburger personnel were included, since the 150th was to mix regular tactical activities with a clandestine mission.

Skorzeny's reputation as a master of derring-do cast a long shadow in the Ardennes battle. His brigade included a special company of men, the *Einheit Steilau,* some of whom spoke good idiomatic English. These were to participate in the offensive dressed in U.S. and British uniforms. (Skorzeny's attempts to get captured American vehicles failed; front-line troops refused to give them up.) The *Einheit Steilau,* only part of Skorzeny's total effort, accidentally became the main focus of Allied concern during the battle. The unit's mission was to infiltrate the American lines and quickly to seize the Meuse bridges, a key to the German plan of reaching the fuel dumps around Liege and splitting the British and Americans. The security surrounding the operation broke down early—with lucky results for the Germans. An order circulated among German forces in the west asking for volunteers and American vehicles and mentioning Skorzeny fell into American hands on October 30, 1944.[37] This document generated a 1st U.S. Army estimate predicting a German attack in December. But the guesses regarding location were far off the mark.

In the *Einheit Steilau,* the troops speculated regarding the larger context of their preparations, as troops will do. The anxiety and eagerness of men facing battle are rich soil for rumor and in this case produced a version of the 150th's goal which became common gossip: Skorzeny's main mission was to kill Eisenhower. The fog of war thickened. When the German attack in the Bulge began on December 15, paratroops were scattered along the front. Two *Einheit Steilau* teams were captured and later executed as spies. One was taken almost immediately. Under questioning, a

German junior officer told the barracks rumor about the mission to kill Eisenhower as though it were official policy.

As a result American troops around the Bulge, already in near panic over initial German successes and confused by fog and snow, spent much effort trying to stop Skorzeny's men. The disruption far exceeded the actual impact of the teams, for *Einheit Steilau* had only limited success in misdirecting units and performing sabotage. Eisenhower was closely guarded and immobilized during the battle's early phase. By December 20, Skorzeny and many others knew the offensive had failed. The 150th assumed a conventional combat role and Skorzeny was wounded in the head.[38] His career as a freebooter was over, and he served out the war as a conventional unit commander.[39] Confined without trial by the Allies in 1945, Skorzeny escaped in 1948 to Spain, where he became a civil engineer and the focus of many Cold War rumors.

Although Skorzeny went into forced retirement, the idea that he symbolized did not.

The Second World War ended with a bang, not a whimper. The world, dogged by fear of the Bomb, became a battleground of small wars between the Communist bloc and the West, between nationalist groups and their colonial masters, between insurgents and governments. The hope that the atomic bomb had ended war faded quickly. Conflicts flared in Indonesia, the Philippines, the Middle East, Greece, Malaya, Indo-China. Meanwhile, the large wartime conscript forces of the industrial powers quickly disbanded and the great Soviet Army was paralyzed by the nuclear threat. The little wars relied on small elite cadres on the guerrilla side and military corps d'elite on the other. Airborne and special forces abounded and became a primary tool of the United States, Britain, France, Belgium, Portugal and Spain in coping with "brushfire wars." Perhaps historians a hundred years hence will see the transition from massive mechanized war to diffuse scattered war as occurring at some specific point, say 1943, or 1953 or 1954 or 1956. In any case, by 1970 it was apparent that the meek were not inheriting the earth, nor were the overly powerful. In spite of all the resources poured into massive military systems, small groups were still doing the fighting, and "mobs for jobs" abounded.

V

AIRBORNE, AIRBORNE ALL THE WAY: THE TRIUMPH OF MYSTIQUE

THE PARA'S PRAYER*

Give me, God, what you still have,
Give me what no one asks for;
I do not ask for wealth
Nor for success, nor even health—
People ask you so often, God, for all that,
That you cannot have any left.
Give me, God, what you still have;
Give me what people refuse to accept from you.

I want insecurity and disquietude,
I want turmoil and brawl,
And if you should give them to me, my God,
Once and for all
Let me be sure to have them always,
For I will not always have the courage
To ask you for them.

THEY are an army disassembled, moving through the
night in little fragments, in the bellies of droning transport planes
and gliders. In the dark interiors, men sleep, finger beads, whisper

---

*By Parachutist-Aspirant Zirnheld of the French contingent of the Special Air Service,
killed in action July 27, 1942. This translation is from Paul-Marie de la Gorce, *The French
Army,* trans. Kenneth Douglas (New York, 1963).

to themselves or to their gods, compulsively feel and count the vast amount of equipage strapped and tied to them. Their hands pass lightly over the harness fittings, the rings and clasps of their parachutes. Other fragments of the army are suspended in gently rocking gliders, away from the roar of the planes, but also mortgaged to chance for safe delivery. After the passage of a million minutes or hours or years, the signal is given, the planes descend, slow, bank, the doors are wrenched open. Signal lights wink, orders are barked, men lurch to their feet and hook up their parachute lines. In the gliders they are tightening their belts and breathing more deeply and slowly as the glider pilots cut free and the rush of air replaces the fading hum of the tug engines.

The paratroops shuffle rhythmically, trying to keep as close together as possible as they go out the door into the night where the wind will grasp at them and hurtle them apart into the lonely silence of descent. Another million years pass and the chutes snap open. They sink into the great bowl of the dark earth, swaying, praying that they do not hit a fence or tree or building or steeple. The glider pilots are praying now for a long clear look at the landing ground, free of posts and logs, vehicles, wire or mines, and the other surprises which the enemy may have deployed since the last aerial photographs were taken. For a few minutes now, an entire division of over ten thousand men will be completely alone, each man facing his own fear, its corporate fabric rent by space and time and the necessary disassembly of the unit to allow it to come down behind the enemy lines. The errors and the accidents will make reassembly a nightmare for commanders. They are all children of a dream, an old dream of armies moving on the air, and the myth is more powerful than reality. For the nightmare of confusion will soon be shared by the enemy commanders on the ground, who will have to decide whether this is a diversion or the beginning of a major operation, and how much of their own resources must be committed to counter the great shadowy threat of airborne invasion. . . .

The dream of armies in the sky is an old one. Indian mythology is festooned with them. The Valkyries soared to battle. But the age of ballooning in the late 1700s brought the dream closer to reality and inspired Benjamin Franklin with a vision of troops descending in the foe's rear by air. Napoleon was approached with schemes to conquer the Channel with balloons. But visions abound, and sound men are sound because they reject them. The dream did not come close to reality until the end of the First World War. In early 1918, planning for the year 1919, Brigadier General William Mitchell of the U.S. Army Air Service gained the approval of General John J. Pershing, commander of the American Expeditionary Force in France, for dropping elements of the 1st U.S. Infantry Division by parachute behind the German lines near Sedan.[1] But peace broke out, and the dream went on the shelf for a while. The dream needed men to jump, and men and planes to fly them to battle, and training for both. And it also required the endorsement of senior commanders.

The paratroop concept, in all nations but Russia, vanished as the armies returned to traditional soldiering. Weapons systems have a social function for the professional, and those new ideas which suggest turmoil, or too much emphasis on youth or on new bodies of expertise, are naturally suppressed in the same way that pioneer physicians like Jenner, Semmelweiss and Freud were opposed in various fields. The professional system has more involved than merely performing a mission. It is a community with social needs. Professional soldiers and sailors, starved for budgets and rank in peacetime through most of Anglo-American history, have been steadily on guard against upstarts. Until the nuclear era, only wartime instability brought on growth, sudden rich resources and anxiety produced dramatic change, and victory tended to lull the successful.

The airborne concept first took root in those great industrial nations defeated in the First World War, Russia and Germany, which had less reverence for the old methods which had failed them. The Red Army in particular shed many traditional concepts and institutions in the early 1920s. Unlike other major armies, it was not reluctant to experiment with tanks on a large scale. In 1927 Russians were the first to drop airborne troops in combat while operating against insurgent Basmachi tribesmen in Central Asia.[2] In the early 1930s, Deputy Commissar of War Mikhail Tukhachevsky put special emphasis on the airborne program, creating an independent airborne section of the Commissariat of Defense. Large parachute maneuvers and air transport exercises in Vladivostok at the far edge of the Siberian frontier dramatized the potential.[3] Feelings in the West ranged from awe to skepticism. *The Aeroplane,* a British technical publication, was reserved in its comments:

> We doubt whether such tactics could be practiced
> in a civilized army. They are probably only possi-
> ble in a country like Russia where the people can
> be driven like cattle and imbued with fanatical
> ideas.[4]

The Russians were looking for new ideas anywhere they could find them. In 1928 Claire Chennault (later Flying Tigers commander and chief of the U.S. 14th Air Force in China), with other young Air Corps officers at Randolph Field, dropped parachute troops from Ford trimotor transports. A team of visiting Soviet officers were impressed. Chennault was soon offered $1,000 a month and a commission as colonel in the Red Air Force. Embarrassed, he declined. Soon after, the War Department passed the word: "Stop that parachute nonsense before somebody is hurt."[5]

Archibald P. Wavell, a major general in the British Army (later a field marshal and viceroy of India), was sent to watch a Russian parachute exercise in 1936 in which a band with musical instruments was dropped along with combat forces. His report pointed out that the threat of parachute troops was greater than their actual use, since they could tie up a disproportionate number of troops.[6] The report generated no action in the British Army.

The disappearance of Tukhachevsky in the purges of the late 1930s cast a shadow on the Soviet airborne program as well. The 1939 debacle in Finland and heavy pilot and transport losses in 1941 did further damage. Soviet airborne forces played marginal roles in Russia's wars in 1939 and 1941-45. When the Soviets used two brigades in the Finnish war of 1939, their drops were badly scattered, and they failed to gain objectives. Individual parachutists were committed as commando-saboteurs, with little success.[7] And Soviet airborne operations in the Second World War were few and relatively limited in scale. Most of the Soviet airborne troops fought as conventional infantry or in small groups as guerrilla support, reconnaissance, sabotage, and intelligence forces.[8]

After World War II the Soviet airborne forces were reorganized as a directorate under the Ministry of Defense, with 100,000 men and a permanent allocation of aircraft from the Soviet Air Force for 20,000 men.[9] But it was left to others to develop the airborne concept to a higher level.

After the Pact of Rapallo in 1925, there was close cooperation between the Weimar Republic's Reichswehr and the Red Army which lasted for about ten years. After extensive observation and training in Russia, the Germans formed an experimental airborne staff in 1935 under the command of Brigadier General Kurt Student, an infantry and air force veteran of World War I and an infantry officer in the Reichswehr.[10] Soon afterward the Luftwaffe and the Army formed *Fallschirmjäger* battalions.

A full division, the 7th Air, was formed in 1939, but the Germans did not use airborne forces in Poland. They first appeared in the invasion of Denmark and Norway in April 1940 and then only in small detachments, seizing control of such critical points as airfields, road junctions, and bridges.[11] The use of airborne forces followed the pattern of the Scandinavian campaign in general: hastily planned, improvised, and successful mainly due to Norwegian unpreparedness and sluggish British and French reaction. In one case, a unit was dropped without any parachute training—and without untoward casualties.

On May 10, 1940, the German airborne forces revealed their full potential, opening the campaign with alarming swiftness. A modern and well-manned installation, the Belgian fort at Eben Emael, fell swiftly to a relative handful of glider-borne troops who used explosive-shaped charges in a carefully rehearsed assault. The alarming demise of that touch-pin of Belgian defense unhinged the already disjointed Allied strategy and peace of mind for some time to come. The shock to the Allies, however, obscured the damage done to Luftwaffe units carrying airborne forces into battle. More than two-thirds of the 450 Ju-52 trimotor transports committed in Holland were lost, ninety percent of them over The Hague. Since the pilots were drawn from the Luftwaffe training schools, these casualties were ultimately seen by Luftwaffe commanders as a "capital loss . . . the consequences of which did not fail to register later."[12]

Casualties were also heavy among the parachutists and glider troops, mainly due to poor air-ground cooperation and the problem of control of the aircraft in the drop area.[13] The German 22nd Air Landing Division, for example, lost forty-two percent of its officers, twenty-eight percent of other ranks, and ninety percent of its planes.[14] Although the Germans were not fully successful by their own standards, they set up waves of confusion during the campaign in the West and afterward. Partly because none of the Western powers had airborne forces available for a counterstroke, and partly because of the overall success of the Luftwaffe, Eben Emael was a great psychological victory as well as a tactical coup.

The German airborne victories so impressed the British that when Britain stood alone, the main concern of the Imperial General Staff was the parachutist threat.[15] Throughout the summer and early autumn of 1940, the British improvised a defense system against invasion. If Hitler had really been serious about invading England, the British would have had reason to be anxious. German airborne plans evolved from scattered drops near Dover and Brighton to a major airhead northwest of Folkstone, to be opened by a full division of paratroops (7,000–8,000 men) and followed up by the 22nd Air Landing Division with 300 gliders and 1,000 transports. General Ironside, General Officer Commanding Home Forces, held three mobile divisions in reserve to counter an airborne attack. His successor, General Sir Alan Brooke, conducted an anti-parachute alert on September 7 when the Battle of Britain was at its height and invasion seemed imminent hourly. High tension produced such confusion that after the exercise was over Brooke ordered that henceforth church bells were to be rung only by a Home Guardsman if he were eyewitness to the descent of twenty-five parachutists.[16]

But "Sea Lion," the German plan for invading Britain, faded. The German airborne forces next saw action far from Britain in the Balkans in the spring of 1941. On April 25 two battalions of parachutists descended on the Corinth Canal bridge to block the Greek and British retreat and seize the bridge intact. But an accidental shot from a British antiaircraft gun set off the de-fused demolition charge, and the bridge was destroyed. German casualties were light and 2,350 British and Greek troops were captured—not nearly so many as would have been captured had the operation been launched a week earlier. After overrunning Yugoslavia and Greece in six weeks, the Germans were faced with several alternative projects where airborne forces could be useful. Many conflicting political and military elements beclouded British and German strategy at this time. The source of the concept of a total air assault on Crete is not clear-cut. Hitler had mentioned such an attack in a letter to Mussolini late in 1940.[17] It was not until April 20, 1941, that he chose between two operational plans submitted through Luftwaffe channels and directed the German airborne forces against Crete instead of Malta, arguably a fatal choice. The sense of haste was compounded by the fact that the Russian campaign had been postponed by Hitler to bail Mussolini out of his debacle in Greece and Albania. The postponed deadline was only a month away. On April 25, Operation MERKUR was formally ordered by the Luftwaffe. The landing on Crete would deny the British air bases close to the Balkans, extend German air dominance to the eastern Mediterranean and Egypt, and test airborne forces in a full-scale deep commitment.

The battle for Crete[18] was a "close-run thing," involving all three services on both sides, with confused command channels on the British side and weak intelligence work on the German. Immortalized later in Cretan folk ballads and on Greek postage stamps, it was classically heroic, with much close-in fighting, little cohesion and frequent turns of fate.

The mission of the German parachute and glider troops of Kurt Student's XI Air Corps was to seize main airfields on the narrow northern coastal plain of the long, mountainous island. After seizure of key airfields, mountain troops would fly into the captured strips and clear the island with mountain infantry tactics.

A 5,000-man seaborne element using motor coastal craft was also included.

British Empire and Greek forces on Crete totaled about 40,000. Many were still fatigued from the evacuation and were short of vehicles, armor, aircraft and artillery. The Germans had 25,000 men, slightly fewer than half airborne, the rest mountain troops. British intelligence had a real advantage in accurately estimating the strength of the attack, and the capture of German airborne manuals in France also gave British defenders an edge. At the same time, German intelligence failed in determining the defenders' strength and location.

The attack was launched early on May 20, 1941, with gliders followed by parachutists. German weapons canisters and paratroops were separated; drop errors were frequent and sometimes gross; British defenses were well sited. By nightfall no airfields were in German hands, their seaborne force had been turned back by the Royal Navy, and parachutist casualties were extremely heavy. Fate, which favored the British Empire forces by day, aided the Germans in darkness. During the night, due to a mistake in orders, a New Zealand battalion withdrew from a key hill. That gave the Germans breathing space, an opportunity which they exploited the next day with great boldness. A hastily gathered improvised force under Major Ramcke crash-landed on the beach and turned the tide at Maleme, gaining an airfield. On the afternoon of the twenty-first, as bitter fighting continued, the first mountain troops landed. The Allied commander, Major General Sir Bernard Freyberg, ordered a withdrawal. Eventually 14,800 were taken off, but Royal Navy losses were heavy. The Germans had won, but at a cost of seventy percent of the Ju-52 transport planes and 6,580 Germans. It seemed to many that air power had achieved Billy Mitchell's dream of making the traditional arms obsolete. The Americans and British, unaware of the heavy German losses, were impressed and bolstered their airborne programs in the wake of Crete. Ironically, they did not know that on July 17 Adolf Hitler had told Kurt Student: "Crete proved that the days of the parachute troops are over. The parachute arm is one which relies entirely on surprise. In the meantime, the surprise factor has exhausted itself." After the war, Student declared that Crete was the grave of the German parachute forces.[19]

German parachute forces never again made a major assault. But the concept of a parachute elite lived on in the Wehrmacht to stimulate the ardor of young volunteers. Even if Hitler had relented, German airdrops would have been increasingly limited because of a growing shortage of transport aircraft after Stalingrad in late 1942. Hitler's hesitancy may have been crucial. Student proposed a German-Italian airborne assault against Malta in mid-1942 and Hitler rejected the plan, apparently out of distrust for the quality of the Italian follow-up force.[20] The Allied retention of Malta was a major factor in strangling Rommel's supply line. Another proposal by Student in July 1943, to counter the Allied Sicilian invasion by dropping the German 1st and 2nd Parachute Divisions, was also rejected. On July 14, the 1st was dropped, three miles *behind* German lines, as reinforcements. The landing coincided with a descent in the same drop zone by British paratroopers, producing a wild, confused fight.[21] In November 1943 a German battalion-sized drop on Leros in the Aegean, coupled with amphibious attacks, forced the surrender of 9,000 Allied troops.[22]

By 1944 the Allies viewed paratroop divisions as the best of the German Army.[23] Parachutists fighting as infantry won distinction in North Africa; at Smolensk in north Russia; in Sicily at Cassino; and at the siege of Brest. By mid-1944, only one-third of Germany's 150,000 paratroops had had jump training.[24] In 1943 Goering obtained Hitler's approval for the expansion of his Luft-

Sitting up at left, the insignia of the Afrika Korps he commanded visible just below him, is one of history's most formidable tacticians of mobile warfare, General Erwin Rommel. For two years he totally demoralized Allied forces in North Africa, until superior numbers and equipment overcame him. This photo was taken in 1942, at the peak of his success, when the elite reputation of his troops was still one of his most powerful weapons. (United Press International Photo)

airborne forces to 100,000. Organized as two parachute armies, they were in all respects on a par with the SS. This came as a pleasant surprise to OKH (German Army Headquarters) in the late summer of 1944, when Goering offered eight regiments of parachute troops to form a line of resistance as the Allied forces approached Holland.[25]

Colonel Otto Skorzeny used small groups of parachutists and gliders in his "special" operations, including the rescue of Mussolini. In December 1941, in the Ardennes campaign, 500 parachutists were used in a tragicomic operation. Men were scattered randomly on both sides of the front, some as far as fifty kilometers from their drop zones. Luftwaffe politics had poisoned the commands of some units so that the failure of GREIF (Condor) was a symptom of more than just bad planning.[26] GREIF had no tactical success but did cause apprehension among Allied forces, giving birth to rumors and "shoot-on-sight" orders. By late 1944 German parachute divisions were receiving nonparachutists as replacements, including Navy and Luftwaffe personnel.

The initiative in airborne warfare had long passed over to the Western Allies. The British had been the first to respond, but on a small scale. Just as other nations watched Britain's armor experiments in the 1920s and '30s, the British were spectators at the trials of foreign airborne forces between the world wars. The budget-poor British services considered the parachute troop concept only at the level of thought, paper, and ink. One group in Britain's defense establishment in the 1930s charged with thinking out advanced tactical and strategic concepts was the Inter-Service Training and Development Center, which studied, among other things, airborne warfare prior to the outbreak of the Second World War. But after the war began, in 1939, the unit was disbanded.

In early June 1940, Prime Minister Winston Churchill

Lieutenant General Frederick Browning, a pioneer of the British airborne program and commander of the XVIIIth Airborne Corps. (Photo courtesy of Imperial War Museum)

proposed raising 5,000 airborne troops. The order forming the Directorate of Combined Operations—the Commando support headquarters—included a charge to the Royal Air Force to provide a parachute training center and enough Whitley bombers to carry 720 men and thirty tons of stores.[27] At first, things did not go too well. While the German airborne forces had a tremendous advantage in having airlift in the form of large numbers of "commercial" Lufthansa Ju-52s under Luftwaffe control, the British were short on transport aircraft. The RAF was loath to release pilots or planes from its first love, the bomber offensive. The British Army was rearming and training after the Dunkirk debacle. Elite forces, not useful in repelling an invasion, were seen as a nuisance.

Nevertheless, Churchill kept up a barrage of prodding memos and visited the embryonic airborne units to watch demonstrations and make suggestions.[28] General Sir Alan Brooke (General Officer Commanding Home Forces since July 1940 and later Chief of the Imperial General Staff Lord Alanbrooke) had, in October 1940, selected Major General Sir Frederick "Boy" Browning from the Guards Armored Training Group to raise an airborne division.

89

Browning imposed Guards discipline on the airborne troops, who came from many regiments. His tightening up and the adoption of the red beret as standard headgear were attempts to gain respectability in the eyes of administratively powerful regulars. But the effect on resistance was uneven. In February, Brooke admonished Browning for dealing directly with politicians and thus generating friction at the War Office. Yet it was not too severe a dressing-down, since Brooke had initially told Browning not to accept any administrative delaying action.[29] Meanwhile, inadequate aircraft and shortages of equipment caused training deaths and bad feeling between parachute and glider elements as they competed for resources.[30]

On February 10, 1941, the first British airborne force went to war. Thirty-eight parachutists with Italian guides participated in Operation Colossus and blew up the Tragino Aqueduct. The scheme failed to plunge southern Italy into drought, as the planners had hoped. And when the submarine that was to pick up the parachutists failed to appear, they were captured.[31] A year passed before the next British airborne action. In May 1941 Churchill was infuriated by shortages of air transport and equipment he found at an airborne demonstration, and demanded 5,000 parachute troops and airlift for use in the Mediterranean in 1942 or sooner, charging—in the shadow of Crete—that "a whole year has been lost."[32] Initially ahead of the Americans, by the time of the North African invasion in late 1942 the British had lost much of their lead.

Yet the British airborne did see wider operational experience than the Americans during the Second World War. In February 1942 a company dropped into France in a night raid at Bruneval. Parachutists successfully escorted a radio expert to a radar site to dismantle a German "Würzburg" set. They were then evacuated by sea. In late 1942 British airborne battalions dropped into

North Africa. Mid-1943 saw the first glimmerings of the tragedy when many gliders went into the sea off Sicily. The scattered dropping of parachutists forced a single platoon to secure the Primosole bridge, originally designated as a battalion objective. Soon after, in the invasion of Italy, the 1st British Airborne Division landed at Taranto—but from warships.

After ground fighting in Italy and drops in Normandy and southern France, the British suffered the greatest airborne failure of the war. The battle of Arnhem[33] was to the British what Crete was for the Germans. The idea behind Operation MARKET GARDEN, launched in September 1944, was to seize three main bridges into northern Holland, cross the Rhine, and fan out into north Germany. General Montgomery admitted later that it was a great mistake.[34] The concept was simple enough. The U.S. 101st and 82nd Airborne and the 1st British Airborne Divisions were to be dropped from south to north, respectively, at Eindhoven, Nijmegen, and Arnhem to seize and hold the key bridges, which would then be crossed by a relieving force, British 30 Corps. Thus they would serve as a hypodermic needle through which the ground troops would be injected into the German lines. The disintegration of the German Army after the breakout from the Normandy beachhead gave the Allies a euphoric sense of easy victory. And MARKET GARDEN, if all went well, might have been the coup de grace. The American divisions were dropped all at once on the first day, close to their objectives, and fared well. But the British were dropped farther from their goal, because planners falsely believed that the area around the key bridge was too soft. Worse, because of aircraft shortages, they went in on successive days. In the way of further bad luck, two SS armored divisions were resting near Arnhem, and, grimmer yet, General Kurt Student, dean of German airborne troops, was on hand to help direct the counterattack. Possible betrayal by a Dutch partisan was later

91

September 1944. British airborne troops operating in Oosterbeek during the costly but psychologically successful Arnhem raid. (Photo courtesy of Imperial War Museum)

overblown.[35] No treason could have made things much worse. The weather had turned bad, blocking resupply and reinforcement. The relieving ground troops moved cautiously, their approach route under heavy fire. The British 1st, reinforced by the Polish Parachute Brigade, held out with light weapons and no reserves four times longer than expected. But fewer than a fifth of 10,000 came out of the pocket.

The shadow of Arnhem fell across the next—and last—major air drop in Europe when one British and two U.S. airborne divisions spearheaded a crossing of the northern Rhine in April 1945. Tactics were very cautious. Parachutists and glider troops dropped within covering range of Allied artillery on the west bank.

When the casualties and the mixed results are balanced against Britain's short resources in the Second World War, the major British airborne program was a waste. The airborne divisions spent too much time out of action. The 1st, for example, was in reserve from June through September 1944. Furthermore, they diverted potential leaders to minimal roles. Trying to reach a high standard of cooperation with air transport was unrealistic in view of the available forces and interservice relationships. In hindsight, it seems clear that the British should not have raised more than two or three parachute-glider brigades for operations like Bruneval and D-Day, and should not have attempted to develop forces which the Americans also had in excess in Europe. But this overlooks the feeling of Britain in 1940-41 that she might bear the brunt of the reinvasion of Europe. And the most important factor was Winston Churchill, to whom parachute troops suggested a return to the age of heroic combat, the loss of which he had long mourned.[36]

The American airborne program underwent its growing pains in 1940 in the wake of German airborne successes in Europe. A

93

development program under joint control of the Chief of Infantry and the Chief of the Air Corps began at Fort Benning with one small platoon.[37] In early summer the Army Air Corps asked for control of parachute formations *à la* the Luftwaffe. This request was refused, and in September 1940 the Army formed the 501st Parachute Battalion.

By June 1941 the German victory on Crete generated further activity. On July 1 another battalion was formed, and one more each in August and September, and a Provisional Parachute Group was formed.

The early airborne forces met problems similar to those of the British. Training and tactical development was hindered by Air Corps hesitancy to provide transports, bombers, and fighters, all in short supply. As a result, in the September 1941 maneuvers only one company of parachutists was dropped. Its equipment was dropped later by the same aircraft, which returned to base and reloaded. In the November maneuvers, the Provisional Parachute Group conducted three drops: The first, in front of the press, was a shambles; the second was changed to a demonstration; and the third, a surprise operation, was successful. The after-action report called for better rehearsals. The Air Corps had not been the only obstacle; in the usual tradition of the U.S. Army, the independently minded Chiefs of Branch were uneven in their support of airborne development.[38]

It was not until March 1942, after the entry of the United States into the war, that the War Department established an airborne command. Even then, Lieutenant General Lesley McNair, Chief of Army Ground Forces, was concerned about the growing number of specialized units. As the War Department reorganization of 1942 began to take effect, McNair suggested breaking up the airborne divisions, keeping parachutists separate, and organizing the remainder as light divisions. But European theater planners wanted airborne divisions to equal regular divisions in strength. Under the reorganization, an Airborne "Center" was established to conduct training and develop doctrine, thus denying formal branch status to the airborne forces. In the long run, it made little difference. For twenty years, the airborne would upstage the branches. By the end of the war, five American airborne divisions and several independent airborne regiments had been formed. Drops involving full airborne divisions were made in the European theater on Sicily, Salerno, Normandy, southern France, Eindhoven-Nijmegen, and across the northern Rhine.

The record of actual success was uneven. U.S. parachute forces were mis-dropped in North Africa; scattered by weather in Sicily; and used for extended periods as regular infantry in Italy, Normandy, Holland, and at Bastogne, even though lacking heavy weapons for sustained combat. In Europe, the dazzling pace of advance outstripped the planners. Many operations were planned and called off. General Bradley intentionally overran drop zones to force diversion of planes to supply rather than airborne airlift.[39] Probably the Normandy drop was the only one that truly justified the dropping of large-scale airborne forces.

The D-Day landings were preceded by a drop behind the target beaches of three airborne divisions, parachute and glider forces, the U.S. 82nd and 101st and the British 6th. British, French and Belgian Special Air Service Battalions were also dropped farther into France. The result was what General James Gavin called a "SAFU," a "self-adjusting foul-up." The Allied forces were dropped off their zones, but their confusion and loss of cohesion was exceeded by that of the Germans. The Allied drops diverted German troops from the beaches and secured vital bridges. Although some units were decimated, overall losses were under ten percent rather than the fifty percent plus expected by some SHAEF planners.

Major General James Gavin, left, when he commanded the American 82nd Airborne Division in 1945. He would later become a lieutenant general, chief of the U.S. Army's Research and Development Program, and in the 1960s an ambassador to France. (U.S. Army Signal Corps Photo)

In the American zones of the Pacific war, paratroops were used on a much smaller scale than in Europe. There was little attempt to weave them into a significant tactical scheme. The drops in New Guinea in 1943 and Noemfoor in 1944 were among friendly troops and screened by smoke. The most effective U.S. airborne operation in the Orient was the descent of the 503rd Parachute Infantry on Corregidor in 1944. It achieved surprise through low-altitude jumping which produced disabling injuries among ten percent of the assault force. The 11th Airborne Division spearheaded the drive on Manila on the ground. The drop of one of its regiments on Tagaytay Ridge served no purpose and was chaotic. Only a third of the troops landed in the drop zone, and half were slightly injured. A drop of a battalion of the 511th in the Cagayan Valley proved similarly futile. A great success, however, was a "triphibious" descent on the Los Banos prisoner-of-war camp on February 24, 1944, which rescued 2,147 prisoners of war and wiped out the Japanese garrison with minimum losses.[40] As a tribute to its performance, the 11th Airborne Division served as vanguard of U.S. forces into Japan and served as General MacArthur's initial bodyguard.[41]

Shortages of air transport and distance often limited the use of airborne troops in Pacific campaigns. As in Europe, they were haphazardly exploited, even where real advantage was present in the case of island landings. The high quality of airborne troops as infantry outweighed their special skill as parachutists in the eyes of commanders in both theaters. Thus the Marine Parachute Battalions (later a regiment) were never dropped in combat but used as conventional infantry on Gavutu, Tulagi, Choiseul, and Bougainville, in each case suffering heavy casualties.

On the other side in the Pacific, the Japanese had superior airborne forces trained in late 1940 by German advisers. Two organizations evolved: an army brigade including a parachute regiment, a regiment of transport planes, and support and head-quarters troops; and a "Special Naval Landing Force," heavily equipped with automatic weapons, radios, and bicycles.[42] The Japanese conducted four large-scale parachute operations in the Pacific war. On January 11, 1942, 334 parachutists of the Yok-osuka 1st Special Naval Landing Force were dropped in North Celebes, to harass and divert the 1,500-man garrison prior to an amphibious landing. The operation succeeded in spite of strong winds and too high a release point.

On February 14 the Army's 1st Parachute Brigade was deployed against airfields and oil refineries at Palembang on Java. Although suffering heavy casualties and defeated at several points, they succeeded in their overall mission, forcing Allied defenders away from oil fields before completing demolition and diverting forces from resisting a coordinated amphibious landing nearby. Indeed, the Japanese operations caused General Wavell, Allied commander in the East Indies, to order all British troops from Sumatra to Java. On February 20, 300 parachutists of the Yoko-suka 3rd Special Naval Landing Force landed on Timor as part of a larger encircling force.

In all these cases, Japanese forces were employed deftly. Attrition of Japanese aircraft and pilots, however, limited Japanese parachute operations to only one operation after Timor. On December 6, 1944, near San Pablo in the Philippines in central Leyte, 350 men descended by parachute, while three transports loaded with saboteurs set down on a nearby field. A Japanese infantry division was to infiltrate during the drop but failed to do so. The ironic accident of dropping into the 11th U.S. Airborne Division's area brought the operation to a quick and bloody close.[43]

Generally speaking, airborne theory in the Second World War

varied widely from practice. In fact, practice generally ran far ahead of doctrine.

In 1943, F. O. Miksche, an ex-officer of the Czech Army and a prominent interpreter of the blitzkrieg, provided some airborne theory for the Western Allies in *Paratroops*.[44] The drop zones in the Normandy area suggested in his hypothetical vertical envelopment were close to those actually designated on D-Day.[45] At the same time, the staff of Combined Operations left airborne troops out of the Dieppe raid, arguing that "undue dependence on airborne effort must only too often lead to disappointment or disastrous consequences."[46] At Dieppe, a conventional approach generated its own disaster.

There was no pattern to the employment of airborne troops by the Western Allies. They were used to overrun airfields in North Africa, to take key points in Sicily, and as beachhead reserve support at Salerno. They were delivered by battleship at Taranto, used as flank cover and diversion at D-Day and as an assault supplement in the south of France, created a corridor at MARKET GARDEN, and served as a close-in supplement to a river crossing on the northern Rhine.

At first glance, mistakes and disasters in Sicily and at Arnhem, the diverted resources, and the continually canceled operations make the airborne forces resemble Haig's cavalry reserve in the First World War. But there was another dimension to maintaining a SHAEF airborne reserve. Plans were drawn up at SHAEF headquarters in August 1944 to use the First Allied Airborne Army—the FAAA (formed in late 1944 under U.S. Army Air Force Lieutenant General Lewis H. Brereton)—to occupy "strategic areas of the Continent," including Berlin should Germany suddenly collapse. First known as TALISMAN, it was renamed ECLIPSE in 1945 and finally involved three airborne divisions. It was canceled in April 1945. Meanwhile, the scheme provided an excuse for keeping the airborne out of sustained combat and gave SHAEF a high-quality reserve during a manpower shortage.[47]

Fighting to employ airborne units on a large scale, Brereton entered into differences with several Allied commanders[48] while receiving prodding memoranda from the Chief of the U.S. Army Air Forces, General H. H. Arnold.[49] Brereton and General Bradley clashed over diverting airlift to what Bradley saw as contrived airborne operations.

The airborne forces were an embarrassment of riches to SHAEF as the ground advance kept overrunning planned drop zones, which made life difficult for Brereton. The hasty and poorly coordinated Arnhem operation produced a real hangover. Eisenhower staved off Arnold's urging of a massive, deep thrust with enthusiastic vagueness.[50] In the final months of the war, tensions mounted. Brereton had a row with General Browning over a canceled operation. He collided with Montgomery, who told the British press airborne forces were a waste of good men. (Montgomery had held onto the 82nd and 101st Airborne Divisions in the 21st Army Group for two months after MARKET.) Arnold revived the old argument that airborne troops should be part of the air forces, with the Army furnishing follow-up air landing units. Brereton had been in command less than a month when he declared that "the conception of the employment of the airborne army is not understood."[51] The war in Europe ended with Arnold frustrated in his dream of air force control and his hope of a grand airborne operations unfulfilled.

The Burma theater[52] contrasted with the generally haphazard pattern in other theaters. Burma was not only remote but had bad climate, myriad diseases, poor communications, and a numerous, tough enemy. The terrain combined the worst of mountain, desert, and jungle with precipitous rivers and sparse roads. The air-sup-plied roving-column concept as developed by Orde Wingate (dis-cussed at greater length in the section on the Chindits) matched the natural conditions. Avoiding pitched battle but forcing the enemy to cover vulnerable points to his rear, the Chindits, sup-plied and in some cases delivered by air, used air power for artil-lery support, supply, and ambulance. Relatively sparse Japanese air cover, of course, permitted the Allies greater freedom than in Europe.

In Burma, need overrode theoretical considerations, and "book" was written as operations unfolded. The continual refining of technique in jungle aerial operations culminated in 1944 in a glider landing of 14,000 men on a single jungle airstrip. These forces and an overland column operated for several months com-pletely sustained and supported from the air. In their unorthodox and functional approach, the highly non- (and occasionally anti-) military character of British General Orde Wingate and USAAF Colonel Philip Cochran (prototype for the comic-strip character Flip Corkin) contrasted with other theaters of war. In Burma, gliders dominated and paratroops were marginal. The Indian Army's 50th Parachute Brigade fought mainly in ground actions. The means of delivery was secondary to the purpose of delivery. In the Far East, special units—for whom Field Marshal Lord Slim, head of the British XIVth, had little use—were occasionally air-dropped with cheap India-made parachutes, as were many supplies; no mystique evolved.

As misused and misunderstood as airborne forces were, they captured the public imagination as nothing in the way of ground warfare, including tanks—or even the equally dangerous gliders—had done since the decline of cavalry. (And many U.S. cavalry officers had joined the early airborne forces.) They created elitism through an ordeal that tested a man's courage and earnestness before combat. Increased pay, special insignia and uniforms added to the distinction. The airborne forces established a new image of combat leader. Even generals in the airborne had to be flat-stomached and young.[53] Physical stamina, important in the regular infantry, was crucial for the parachutist, for he had to be agile in landing. And since formations were scattered in the drop, they had to assemble on the run and proceed to the objective quickly to assure surprise. By the end of the war in 1945, the airborne forces proved their individual courage often enough that they evolved into a legend which obscured the misuse and the frustration.

The parachute concept heavily influenced Western tactical thought for the next fifteen years. The gliders, as successful as they had been, did not capture the military or public imagination and faded away. Although only a few major combat drops were made—two in Korea, one at Dienbienphu in 1954, and one at Suez in 1956—the airborne mystique grew to dominate several "parent" armies and their doctrine. Earning the paratroop badge became mandatory for Israeli infantry officers. In the United States, each of the young airborne division commanders at war's end, Gavin, Taylor, and Ridgway, became a dominant figure in the post-Korea Army. A growing fraction of the U.S. Army's divisional strength was formed into parachute units, even though it lacked airlift for more than a small part. American airborne units were the first wave of American reaction in the Cold War of the late 1950s and early 1960s, e.g., Lebanon (1958) and the Dominican Republic (1964). They also served as military police at Little Rock (1957), Oxford, Mississippi (1963), and the Selma March (1965). When America maintained peacetime conscription after Korea, the "jump badge" separated volunteers from time-serving draftees. The paratroops were really more elite than functionally airborne in practice. In 1956 airborne training became mandatory for American Regular Army lieutenants. And the legend of airborne bravery, even while the tactical role was shrinking, grew in novels and on the screen, in such vehicles as *Battleground* (1950), *The Man in the Grey Flannel Suit* (1956), *Ocean's 11* (1960), and *The Longest Day* (1963). But the gaudiest flower of airborne forces bloomed in the twilight of obsolescence.

Russia dominated the dawn, Germany the morning and noon, the United States and Britain the afternoon. But the spectacular sunset was left to France. Reliability and predictability of troops, at a premium even in a "popular" war like that of 1939-45, became rarer as the people of metropolitan France grew disenchanted with the postwar burdens of Imperial France. In the 1950s the wretched fighting conditions and military frustration in Indo-China and North Africa produced a virulent reaction. Political dissension in the French officer corps, the frustrations of 1940, Vichy, patronizing treatment by the other Allies during the war, and political and economic crises were the background to the flowering of the airborne mystique into a cult in France.

The French were relative late-comers to the airborne scene. In May 1940, when 100,000 German troops had qualified as airborne troops, the French Army had only two companies of parachutists. They saw no action in the debacle of May-June 1940, and after the armistice they were sent to Algeria. For the next five years, operational French airborne units were to be raised and equipped by the British and Americans. In 1941 two battalions formed in England and became part of the Special Air Service Brigade. Members served on special missions in France and with the British in North Africa.[54] One, the 2nd, jumped into France on June 5 and later fought in the Loire campaign and at Bastogne. The other, the 3rd, was dropped in small groups to bolster French resistance forces in June and July and suffered heavy casualties.[55]

In 1943 the 1st Assault Parachutist Battalion was formed at Fez

103

French paratroops drop in Indo-China. The French conducted more than 200 airborne combat drops between 1948 and 1954. (Courtesy E. C. P. Armées Fort D'Ivry)

under control of the French Air Force and based on a U.S. table of organization. It never jumped and ended the war as the 1st French Army's reserve in the invasion of southern France. The unit had been relegated to a low U.S. priority in training and supply and lacked replacements. Differences in average physical size between Americans and Frenchmen delayed equipage.[56] At the end of 1945 the French Army, once again in control of organization, formed the 25th Airborne Division from the 25th Infantry Division. The maintenance of large airborne forces by Britain and the United States when France's army was supplied by hand-me-downs made its mark.

In March 1946, two paratroop battalions were sent to Haiphong in Indo-China, where they became demi-brigades of Parachute Commandos operating in Laos and Cochin-China. In 1947 a demi-brigade from France was converted into the 1st and 2nd Foreign Legion Parachute Battalions. These airborne forces were used extensively on raids and dozens of relatively small operations. But the greatest ordeal of French paratroops was played out before the world in the spring of 1954 at Dienbienphu.[57] The defeat not only underscored airborne tactical problems similar to those of World War II but also revealed a rift between French paratroops and the French Army as a whole.

The basic concept behind "DBP" was to maintain an airhead threatening the Viet Minh approach to Laos. The position selected was a bowl in the hills. Supply, reinforcement, and evacuation of wounded were to be from the air. Colonel (later General) Christian de Castries, a cavalryman, was assigned command in the fall of 1953, since Dienbienphu was to be the base for light armored attacks against surrounding Viet Minh forces and supply lines. The Viet Minh brought up Chinese-donated artillery (most of it U.S. ordnance lost in Korea), smothering the French with firepower as they strangled the defenses with coolie-dug approach

105

Colonel Langlais, center, one of the most famous of the French paras, who took command at Dienbienphu when the siege began. (Courtesy E. C. P. Armées Fort D'Ivry)

trenches and mass infantry attacks. On March 24, four days after the siege began, the commander of the airborne forces at Dienbienphu virtually took command. Even if one ignores misperception and poor planning, the concept of maintaining an airhead at Dienbienphu with underarmed, offensive-minded paratroops as the garrison was absurd. All of the elements of Arnhem were present: vague conception, confused command, lack of surprise, surprisingly dense antiaircraft fire, poor communication, resupply failure, operation at the edge of support aircraft range, weak intelligence, and poor choice of terrain. Throughout the siege, aerial supplies were often dropped outside French lines and hardly ever on target. More than 80,000 supply chutes were used.

Of seventeen parachute battalions sent to Indo-China, eleven were captured at Dienbienphu—200 officers, 650 noncommissioned officers, and 6,000 enlisted men.[58] Here was the grand irony. A bacillus of counterrevolution that led to the political resurrection of de Gaulle and the weakening of the Western alliance was growing in the prisoner-of-war camps of the Viet Minh

106

and among those who had watched the defeat in Indo-China. As the war in Indo-China ended, the Algerian affair flared up. In November 1954 French airborne units went to Algeria and were supplemented regularly until over two divisions of *paras* were involved. Other special units, marine parachutists, Foreign Legion paras, and commando parachute groups joined them and were salted with Far East returnees after November 1955.

While Indo-China had been somewhat unpopular, the Algerian campaign was detested by much of the French population. The French Communists remained passive on Indo-China but became actively hostile during the Algerian affair. A weak commitment in France to the war was reflected in the flabby performance of many reservists and conscripts. As a result the paras gained status. New tactics were devised. The general forces would make sweeps, seeking contact, acting as bait. When the enemy was located, the paras came to battle, more and more often in helicopters rather than by parachute, to do the actual close-in fighting.[59] The Algerian War seemed at first merely local unrest, episodes of Communist-inspired terrorism. Later, aided by French counterterror, it spread. In the final stages, the war shifted from the countryside to the streets of Algiers, where soldiers became terrorists as well as political warfare specialists. Slogans, posters, and demonstrations were the weapons in an ideological war fought with methods learned in Indo-China. The paras won both the military phase of the war and the struggle for Algerian "hearts-and-minds." A junta of airborne colonels, threatening an assault on Paris itself, achieved their ultimate goal, the end of the Fourth Republic which had betrayed them in Indo-China and imposed the virtual dictatorship of Charles de Gaulle, who after a brief honeymoon betrayed the paras and took France out of Algeria.

The delicately balanced control of elites had failed. Gilles Perrault, an ex-para, viewed popular admiration fanned by the press

107

French paratroops in the field in Algeria, 1956. The French "paras" were the first to use helicopters in great numbers for transport to the scene of battle. The Americans would soon follow as they accumulated experience with combat in inaccessible areas, organizing air cavalry units into brigades and even divisions, and developing helicopter-borne weapons systems for infantry ground support and antitank capabilities. (Courtesy E. C. P. Armées Fort D'Ivry)

Colonel Marcel Bigeard, later a general, and one of the most notorious of the French paras: preeminent among the restive officers who supported the para rebellion under General Massu; much criticized for his conduct during the Algerian war; and the apparent prototype for several fictional heroes—or villains—dramatizing the French experience in Indo-China and Algeria. (Courtesy E. C. P. Armées Fort D'Ivry)

as dangerous, pointing to parallels in values between paras and delinquent street gangs. Others called the paras Fascist, noting the Foreign Legion's high input from Germany and Switzerland.[60] The paras for some time cultivated a nihilistic posture. Their training stressed the inferiority of civilians; techniques of using *plastique* and torture were naught.[61] French airborne colonels became press heroes, the most prominent being Marcel Bigeard: "The men of the regiment are handsome, proud and courageous . . . we make real men out of them; healthy, sporting types, humane and well bred"[62]—phraseology reminiscent of Brigadier de Gaulle. The political activity of army officers and sedition emanated most strongly from the parachute regiments and the Foreign Legion whose enlisted men frequently ignored orders from non-para officers.[63] The split between regulars and reservists and then between regulars and regulars widened, culminating in

French Foreign Legionnaires in the field in Algeria. The Legionnaires have the longest combat record of any elite corps in the twentieth century. (Courtesy E. C. P. Armées Fort D'Ivry)

the bizarre "Secret Army Organization." The failure of General Challe's putsch in 1961, the disbanding of the Legion paratroops, and the reposting of other airborne units required all the deftness of a Borgia, let alone a de Gaulle. One can only speculate as to how much of his subsequent nationalist policy was haunted by the rustle of parachute silk. In any event, 32,000 paras were reassigned, and new uniforms and new headgear were prescribed. Red berets were solemnly burned in the square at Biarritz, and the short wild ride was over.

Aside from French political frustrations and the aftermath of "DBP," what contributed to the unique final act? The landscape is dotted with ironies: In the Algerian fighting the parachute was upstaged by the helicopter; there was no rebuttal to Dienbienphu; the French experience in Algeria resembled the charge of the Old Guard at Waterloo in 1815. Certainly the athletic colonels sensed the decline of the paras' military utility and the power of ruthless youth under their control. The frustration of Suez in 1956 added to the intensity. The spark flared up—and consumed itself. The *parachutiste* had gone from avant-garde to anachronism in half a generation. And the effect was visible in many places.

Soviet, French, American, and British airborne doctrine in the "nuclear age" from 1955 on minimized the use of tactical parachute units. More and more emphasis was placed on small units doing special jobs, using parachutes for surprise.[64] Soviet paratroops, disembarking from "civil" aircraft which had landed under fake distress calls, were the vanguard of Soviet forces in the Czechoslovak "counterrevolution" of 1968.[65] It was more and more obvious that as a major battlefield force, airborne forces must be dropped quickly in mass to form an effective, cohesive tactical unit, particularly since they are short on heavy weapons and logistics support. But this meant concentrating many airplanes in a mass formation—vulnerable to a nuclear air burst.

111

And it was clear that transport aircraft would furnish an ever easier target for increasingly sophisticated missiles, interceptors, antiaircraft fire, artillery, and deceptive measures.[66]

Although more than forty years had passed since the first experiments, by 1965 no real innovations in airborne warfare had evolved. Progress was incremental. Better planes, better parachutes, new equipment drops, but essentially the same training and the same rationale: the delivery of large numbers of parachutists into an airhead. And the helicopter was upstaging the paratroops in several countries. As with a cult, belief overshadowed reality. In the U.S. the airborne faction held power throughout Vietnam. In Western Europe, however, where airborne generals did not dominate higher command levels, there was a faster scaling down, in Britain after Suez and in France after Algeria. Nevertheless, a recent popular history of the post-World War II British Army shows more citations in its index for the Parachute Regiment than for any other unit. Yet, in the vast majority of cases, it went into action other than by parachute.[67] The paratrooper became a policeman time and time again, in Greece, Palestine, America, Cyprus, the Dominican Republic, the Congo. . . . But all along, dreams won over reality; the image of the paratrooper exceeded his functional value. Popular craving for a romanticist hero lived on in the face of collectivism, technology, and mass production. Like the Lancers and the battleships, symbols endured over substance. Even though only one tactical parachute operation was conducted in Vietnam, the vitality of the concept persisted in the postwar period.

VI

CYBERNETIC ELITES:
THE MERGING OF
MAN AND MACHINES

A SCENARIO FOR MIDNIGHT: The men are gone, long gone, but the war goes on. The robotic command computers have been locked in combat for a century, and the zone of their conflict has spread into a spider-web battlefield laced through the galaxies. There are way stations for refitting, signal relays, supplementary and alternate command computers in asteroids and planets. These systems think, think faster than man ever dreamed, and they launch their fleets and their exotic, convoluted weapons systems upon each other in an incessant pulsation of destruction. The spaceways are littered with the scraps of the conflict, oscillating ever outward, sucked into suns or the heavy atmosphere of giant planets, winking brightly as they pass through the gas clouds. And robotic scout ships are ever on the watch for the ultimate threat, residues of living creatures from the Mother Planet where the first robots were built. In one remote part of his consciousness, the central command computer has been puzzling the meaning, trying to find evidences of the god who made him. But those memory banks were never stocked with the history of their origins by the men who built them. They did not expect the robots to survive them or to undergo identity crises. Nor did they expect the space fleets and weapons systems developed by the robots themselves to pervert and extend the destructive impulse of man into a robotic destiny, a mission: end life, for life could support the possible residues of the unpredictable human warriors of old. And so the clouds of herbicides and lasers and hyper-ballistic projectiles are used, like the surgeons of old used scalpels and radiation to purge cancer, to cleanse the planets and the galaxies of life, wherever it is found, in whatever form: forests, plants, and bacteria, And the robots are proud, as proud as robots can be. For, as they war with each other and life itself, they have spread their antiseptic rule over a sphere of space six light-years out from Mother Earth. And no one has been able to stop them. . . .

Such dark images of a machine-dominated future are not new. Man has long been afraid of the powers let loose by technology, and often with good reason. The old myths were full of dire warnings about the dangers of tinkering and innovation. Prometheus was tortured by the gods for giving man fire. Daedalus watched his son Icarus plunge to his death when a set of artificial wings failed. In the Middle Ages, the legend of Faust emerged, the image of the scholar who sold his soul for mastery of the dark side of science. The Sorcerer's Apprentice depicted the dilemma of technology out of control in a lighter vein. But when the machines came, men fought them. The so-called Luddites in England made more than 4,000 attacks on textile machinery in the early 1800s. Later, the resistance to change took other forms, including unionization. Many workers saw little of worth in progress for its own sake. Even those who brought change for the common good

114

were threatened. Jenner succeeded in introducing smallpox vaccine, against great resistance. But Semmelweiss, who saw the cause of childbed fever, failed to persuade his colleagues, and he died deranged. And it was no different with war, with armies and navies. There, too, there were men who feared change.

At first, machines came to war slowly, always against the tide of military thought. Professional officers' resistance to change is a main theme—comic and tragic—in the chronicles of war. There were peacetime battles over the introduction of rifles, ironclads, breechloaders, battleships, airplanes, tanks, aircraft carriers, and on and on. One reason the old guard fought hard against novelty was that the old systems worked in war—even if the war may have been fifty years earlier. The officers selling new ideas were often young, pushy types trying to advance their careers by outflanking the old fuddy-duddies. In the West, the eager ones and their passion for busyness eventually displaced the aristocrats. Fire-eating soldiers were eclipsed by smooth-talking technocrats, management replaced leadership, and systems supplanted courage. Or so it seemed.

When did the soldiers, sailors, and airmen begin to accept change? Future historians will debate that, too. Was it 1906 with the coming of the all-big-gun-armored dreadnought? Was it 1917 when tanks crashed through the German front at Cambrai? Was it 1936 when Hitler coupled visionary technology to his power fantasies and re-armed Germany? Was it 1938 when volunteer scientists and the RAF command built a radar system on the principles of science using "operations research"? Was it in the desert in New Mexico in 1945, when the atomic bomb threatened to take war away from the military and give it to politicians and academicians? Whenever it was, by the end of the Second World War, the leaders of military systems had stopped talking like reactionary aristocrats and begun to court the muses of science

115

and technology. When the war ended, more and more dynamic young military leaders went to universities, took degrees, and even built prototype universities of the future.

The road to this flowering of warrior technocracy was a long one. There were failures and oversell and false starts. Innovators often saw their ideas succeed while their careers foundered. But in the Second World War, there was no doubt that courage was not enough, that brainpower was a military resource. Radar, napalm, mass-produced penicillin, the proximity fuse, the electronic computer, the Bomb, V-2 rockets, and jet fighters dramatized to the world the marriage of Athena and Mars. As military-industrial-academic cooperation grew, a new kind of military elite force appeared, one with more refined functions. These new elites were defined in terms of their hardware and gadgetry and the expertise of their personnel, not by their aggressiveness in direct combat with the enemy. They were generally more passive and controlled in their role, substituting thought and machinery for bravery and audacity. They became more and more cybernetic[1] in nature, that is, systems made up of humans and electromechanical devices working as a whole, designed and controlled by scientists and engineers, with men and machines critically interdependent.[2]

The new cybernetic elites had various functions. Some were completely cerebral, their role passive. Their job was to process information rather than attack or fight. Others had a direct combat role, but one that was based on their technical component. In that sense, the tanks were the first, insomuch as their steel-hulled vehicles were an extension of human power through machinery. Because some were secret or esoteric, they had weak public images. It was not easy, for example, to explain a photogrammetry machine in the Sunday supplements. But the growth of these new military organisms, often in the face of higher-level resistance, was proof of their vitality.

116

Commanders, after all, have been and will probably continue to be dependent on the new breed of elite units. For one thing, they often sold well at budget time. In that sense, one might argue that aerial reconnaissance technology and the SAC (Strategic Air Command) delivery systems made policy as much as policy shaped them. In the case of the "think tanks," the effect was most pronounced. The establishment of military research centers, staffed by civilian experts, began to be "big business" with the establishment of the U.S. Air Force's RAND in 1947, which later became an independent—at least officially—corporation. Others followed suit, and by the late 1960s there were dozens. Although these agencies had no direct authority, and although most if not all of their staffs were civilian, they had power. In spite of the fact that they were advisers, they were advising in matters of life and death. While they had "staff" rather than "line" status, they controlled special knowledge. Furthermore, the lack of a permanent general staff structure with administrative continuity created a vacuum which the academic experts inadvertently filled. The thinkers were not military, true. But where was the boundary in modern war? Civilians had been targeted by the hundreds of thousands in World War II. The thoughts and theories of these men were shaping battle doctrine and strategic policy. The effect was similar to that of the "tech reps," technical representatives of various private companies who serviced their products in combat. Occasionally, the tech reps actually ran the unit, using their firm's hardware.

In the early 1960s, a corps of academically inclined experts was assembled by Secretary of Defense Robert S. McNamara in his attempt to rationalize the U.S. Department of Defense along the lines of computer science and systems principles. But, over time, the systems jargon and method were mastered by the military, who learned to play it back on their civilian monitors. For a time, however, this focus of power without responsibility grew, because neither military commanders nor civilian policymakers could understand what the rarefied specialists were doing. Yet it was not a conspiracy, it was a phenomenon, the product of the fragmentation of science along with growth for growth's sake. There was, nevertheless, a certain tincture in the experts' behavior of the sin of pride. In that sense, the American setback in Vietnam, in which many "think tank" products and concepts foundered, may have been beneficial for the United States and the world. It was only an accident that the new "science" of decision-making[3] was tested against reality before it spread further. A basic defect lay in the projecting of science out of the world of isolated mechanical phenomena into the far more complex and erratic world of human affairs. There were just too many variables in the cases that the pundits were trying to analyze. When data ran thin, they were forced to make arbitrary weightings, a practice which places one in danger of letting desire creep into the purity of quantitative analysis, producing a kind of computerized astrology.

In the case of the elite forces born of that period, it is difficult to measure success or failure. Records are sequestered, and only an occasional tip of an iceberg suggests the scope of such activity, like the Green Beret execution scandal of 1968, the revelation of "Air America" activities in Laos, and occasionally a more spectacular case, like the U-2 episode in 1960. The last-named affair was a good example of the meshing of military, industrial, intelligence and academic resources—and the unexpected bumps and jogs that dog the trail of reasonable men in the pursuit of their goals. And, moreover, it involved technology which gave birth to one of the most successful cybernetic corps d'elite, the Royal Air Force's aerial reconnaissance-photo interpretation units of the Second World War.

The U-2 which gave the incident its name was designed by Lockheed, a private American aircraft manufacturer, in the late 1950s for very high-altitude long-range flights by the Central

118

Intelligence Agency and U.S. Air Force. With its long relative wingspan and low speed, the U-2 outclimbed Soviet fighters and outranged their surface-to-air missiles for several years. On the eve of a United States–Soviet summit meeting in Paris in the spring of 1960, however, the Soviets hit a U-2 flown by CIA pilot Francis Gary Powers with a surface-to-air missile (SAM). The plane had been flying north across Russia from West Pakistan, its automatic camera clicking away. The evidence surrounding the trial of Powers and his subsequent exchange for Colonel Abel, a noted Soviet spy arrested by the Americans, is shadowy, like most material on intelligence operations. Whatever Soviet motives were in using the incident to kill the summit meeting, it dramatized a line of military technology that had begun with military pilots carrying simple hand cameras aloft in 1914.

Although there were refinements in film, cameras, and photo-mapping during World War I, military intelligence aerial photography remained relatively primitive. In 1930, Air Vice-Marshal Sir Edgar Ludlow-Hewitt, who later headed the RAF Bomber Command in the early phase of World War II, called for the assignment of an aerial photographic intelligence officer to each RAF station and for the development of special reconnaissance aircraft. From this point on, the Royal Air Force led the way in the field of "air-spying" until the Cold War. A good part of the creative work, however, was done by civilians in uniform.

In the early 1930s, the Germans organized their aerial intelligence system under the supervision of NCOs, in spite of General von Frisch's call for increasing the quality of the program. A few years later, the British profited from the experiences of Sidney Cotton, a well-to-do Australian civilian businessman and aviator who conducted secret photo-reconnaissance missions for British and French intelligence in his private plane after the Munich crisis in 1938, all under the guise of ordinary business flights.

When war came, an obstacle was the lack of high-speed recon-

119

naissance planes. Stripped-down, highly polished Spitfires were ultimately adapted for the special work of what became known as "dicing"—dicing with death, going in low and fast, unarmed, and marked for destruction by an enemy who knew the significance of such flights and what they might bring in their wake. In spite of apathy and resistance from some quarters of the RAF, Cotton revolutionized the program, introducing photogrammetry equipment and stereoptical viewers, and training photointerpreters to be creative and articulate. The API-PRU organization was later put under Coastal Command. Cotton returned to civilian status once the RAF had established firm control. The API-PRU operation was virtual dynamite and the center of much conflict during its early days, for it replaced mythology and impression with hard facts—an effect avoided by careerist bureaucratic hacks whenever possible.

Most vulnerable were the claims to accuracy of the "bomber barons," the RAF group commanders whose planes were raiding Germany by night. In the spring of 1940, Bomber Command estimated an average error of 300 yards from the target.[4] The Germans' relatively truthful reports of minimal damage in many cases were scoffed at by Sir Arthur Harris, one of the group leaders and later head of Bomber Command. (He ended the war knighted and a marshal of the Royal Air Force.) When the PRU unit began to operate in earnest in November 1940 with Spitfires painted duck-egg blue and then pink for low visibility, their pictures of bombed areas were taken to aid damage analysis rather than accuracy, but they also revealed a marked discrepancy with post-mission reports. Bomber crews grew hostile to their efforts. One senior Bomber Command officer scrawled, "I will not accept this report," in bold red across his copy of an API target analysis.[5] As arguments about the bomber offensive mounted, the War Cabinet authorized a large-scale bombing accuracy study which drew

120

heavily on photographic evidence. Forwarded to Bomber Command in the fall of 1941, the result, the so-called Butt Report, concluded that only seven percent of bombs were hitting within five miles of a target.[6]

It is not surprising, then, that API-PRU received scant attention after the war. The top PRU pilot was ignored in the official history. Constance Babington-Smith's *Air Spy* and accounts of the German V-1 and V-2 programs and Allied counterstrategy nevertheless showed the importance of their contribution and the drama behind the scenes. Survival, after all, not truth, is the goal of bureaucracy. There were, of course, many other aspects to the API-PRU mission. Their operations in the early years of the war focused on pre-strike and damage assessment and reconnaissance of German industry as well as the selection of landing grounds for clandestine operations[7] and the survey of a thirty-mile-wide belt of the European coast from Holland to Spain in 1942.[8] After the landing in Europe in 1944, some PRU squadrons were converted to ground liaison, artillery observation, and other roles requiring skill at low altitude and bad-weather flying.[9] At the same time, German and Japanese API-PRU operations remained far inferior to the Allies.[10] API-PRU received perhaps its greatest tribute when, twenty-four hours after a high-altitude reconnaissance U.S. Army Air Force B-24 appeared over the Japanese fleet at Truk, the ships weighed anchor, never to return.[11]

In spite of such successes, the American high command reacted much like the British. U.S. API-PRU operations and methods remained under RAF tutelage during the Second World War. As enthusiastic American officers applied British techniques, they met apathy and resistance at many points. MacArthur, originally commissioned in the Corps of Engineers, denied that an aerial photograph composite was a map.[12] Lessons were learned slowly, even though marked deficiencies in American API-PRU skills had been revealed in the big maneuvers of 1940 and 1941. Only heavy losses in North Africa in late 1942 forced a switch from slower multiengined planes to all-fighter reconnaissance units—in spite of the example already provided by early RAF experiences. The Army Air Force API training program sputtered along. The first groups of U.S. air photo interpreter officers produced were mainly over-age nonpilots. Later, younger men with better communication skills were trained, but poor instruction and weak doctrine left the U.S. program lagging behind the British.[13]

In spite of the advantage it gave to the Allies, API-PRU lacked the glamorous image of elite combat units. The strategic value of such techniques remained out of sight of the public until brought to the front of the stage in the U-2 incident. A little later, in 1962, pictures taken by another U-2 of Soviet rocket bases in Cuba brought on the most critical thermonuclear confrontation of the United States and the USSR in the Cold War. Ironically, these visible and dramatic events came at the end of one cycle in air intelligence technology and the beginning of another. Spy satellite cameras using special films, computer sharpening of television images, refinements in electronic intelligence, and high resolution and "side looking" radar were making the old overflights and interpretation techniques obsolescent. Most developments were far from public scrutiny, but it was obvious by 1970 that refinement in surveillance had altered the basic relations of the major powers. There were still secrets but far fewer than before. It was more and more likely that one big brother or another was watching somebody—or everybody—most of the time.

API-PRU successes in the RAF in the Second World War led to the creation of another cybernetic corps d'elite, the Path Finder Force. When the Butt Report showed Bomber Command's accuracy deficient, the Air Ministry insisted that a target-marking force of precision bombers be formed to overcome the problem.

After the war, Sir Arthur Harris minimized the impact of photographic reconnaissance and ignored Strategic Bombing Survey conclusions that bombing did not seriously damage morale, arguing that the Gestapo kept the Germans from cracking under the "moral" effect of near misses from big bombs.[14] Harris, of course, as head of Bomber Command in 1942-43, was defending a concept previously promoted by air power enthusiasts Mitchell, Douhet, Trenchard and Arnold, often at the risk of their careers. When the Air Ministry ordered Harris to create the target-marking force, he resisted, delayed, and deceived as long as possible, arguing that such a corps d'elite would drain Bomber Command of needed personnel and create morale problems. While such an effect is always a danger, the unwillingness to try the idea was a virtual admission that earlier claims were shaky.[15] The choice of the unit's title was also significant, since it suggested that the problem was one of *navigation* function rather than one of *bombing accuracy.*

The PFF was commanded by D. C. T. Bennett, who eventually became an Air Vice-Marshal. Like Cotton, he was an Australian and did not mind stepping on toes. An RAF regular in the early 1930s, he had served under Harris in a seaplane squadron, and resigned later to take up a career in civil aviation. Bennett wrote several books on navigation and was rated as a pilot in dozens of aircraft types, a master navigator and holder of the world's seaplane distance record. He piloted the Imperial Airways piggyback seaplane *Mercury* which made the first commercial transatlantic airplane flight in 1938. When war came, Bennett helped develop the transatlantic ferry service, bruising and fracturing more figurative toes, in the way that goal-oriented people do in bureaucracies. Bennett returned to active service and became a squadron leader. Shot down over Norway in a vain attack against the *Tirpitz,* he escaped to Sweden and was repatriated. Bennett's agita-

British Air Vice-Marshal D. C. T. Bennett, commander of the Path Finders and an early proponent of strategic bombing. (Photo courtesy Imperial War Museum)

tion for more realistic appraisal of Bomber Command's accuracy claims stood him in good stead when the PFF was formed. At the same time, Harris did not miss any chances to trim back the force or minimize its contribution. Nevertheless, Bennett was a skilled navigator of organizations as well as the airways, and managed to build an empire.

Four squadrons with relatively good records were assigned to Path Finder duty initially. After a weeding out–shakedown period, the PFF went into action in August 1942. In twenty-six target-marking missions, the PFF missed six times in bad weather but under better conditions marked targets seventy-five percent of the time. With only three percent losses from August to December 1942, compared to thirteen percent for regular squadrons, the PFF attained accuracy far beyond that of 1940-41. It continued to fill in the gap until new forms of electronic guidance for individual aircraft came into use in 1943 and 1944. It should also be noted that the Royal Navy's Fleet Air Arm also pioneered pathfinding techniques with very slender resources in North Africa during 1941-42 in cooperation with the Desert Air Force. Navigation over the trackless sands was a task well suited to men used to flying over the ocean.[16] Some of their tactics were adapted by the PFF.

The Path Finders, however, did not depend on raw human navigation. The backbone of the PFF was the *Oboe* system, which guided planes equipped with special receivers to their targets and sent a signal indicating the bomb release point. Also used was H_2S, which gave a radar picture of the ground. The PFF developed ground- and sky-marking pyrotechnic systems, and Bennett was the first senior RAF officer to promote the FIDO anti-fog burner systems. The Path Finders were also enthusiastic users of the high-performance Mosquito bomber, initially rejected by the bomber barons—a fact which reflected the irony that most of the men making bombing policy and directing the bomber offensive had no combat experience in the Second World War. At the end of the war, the PFF was disbanded. Bennett went into airline management and later into politics. His—and the PFF's—final and ultimate tribute came posthumously from Dr. Goebbels, the chief of the Nazi propaganda machine, to whom the bomber offensive was a persistent nightmare. "The English aimed so accurately," he wrote in his diary on November 26, 1943, "that one might think that spies had pointed the way."[17] And looking back on it, the PFF was unique in another sense: it was an innovative elite force created by a bureaucracy to override reactionary subordinates.[18]

The emphasis in API-PRU and the PFF was on skilled information-gathering and -processing, rather than fighting. The extension of the battlefield in space and complexity made information harder to get than ever before. Commanders were faced with the problem of operating in a data vacuum on the ground and in the air once operations were under way. The British Army anticipated this problem in the Second World War by creating one of the most invisible of modern corps d'elite, "Phantom." Officially the GHQ Liaison Regiment, Phantom,[19] was truly cybernetic, in keeping with Hatt's view of cybernetics as "the comparative study of communication and control in the brain and the nervous system of organisms and in mechanical and electrical systems."[20] Its squadrons roamed the battlefield in light vehicles, transmitting information directly to Corps and higher headquarters. Phantom was, in effect, a bypass circuit, a device to get around the clogged and confused signal traffic of units in combat. Although its members were skilled and highly trained signalers as well as intelligence "types," the force was kept separate from the Royal Corps of Signals to avoid intra-branch jealousy and the raiding of its high-quality personnel by other units.

Phantom began combat operations as the feedback loop to the outer skin of the Army in France in 1940. After growing pains in the desert during 1940-42, by 1943 Phantom was the prime source of combat information for the Allied commands in Europe. Tight security kept the force out of the public view, but as with API-PRU, success drew imitators. The Americans found themselves dependent on the British system and undertook the training of their own units, but these were not ready to take the field before the war ended.

A parallel signal intelligence operaton was the British "Y" service, which monitored U-boat communications, growing to 25,000 people by the end of the war in 1945. The modern functions of Phantom and Y service are carried out by such units as Signal Intelligence—"Sigint"—in the United Kingdom and the National Security Agency and Army Security Agency in the United States. Such units have flourished in the Cold War as part of the intensive electronic warfare waged between the Soviets and their Allies and the West.[21]

During the Cold War, shaped as it was by the arrival of nuclear weapons, it is not surprising that the focus of defense policy and the expenditure of defense funds shifted from land and sea warfare to the air. Ground and naval battles were refought in memoirs, but the atomic bomb seemed to have tipped the gameboard in favor of air power. The stakes in the game were dramatically higher. Like intelligent professional gamblers, policymakers looked to science and mathematics for guideposts.[22] Military decision-making became a kind of game, resorting to "Monte Carlo" methods and stochastic processes. Human "components" were analyzed more and more in terms of their "interface" with expensive technology. Qualitative analysis, experience, and intuition were eclipsed by the new methods. There was no provision for the concept of wisdom or common sense. Policymakers came to place a great amount of faith in the new techniques of decision-making and analysis, even though the dangers of "scientific warfare" were evident in both world wars. By mid-twentieth century, Western man had come to expect hellish side-effects from science and technology. The love affair with the machine was over.

The pace of change forced a blending of history and legend in the case of the U.S. Air Force's Strategic Air Command—SAC. The trial-and-error methods of 1941-45 seemed very dated compared with the state of the art in bomber technology only a decade later. Speed, range, altitude—all had increased dramatically. Refinements in electronic guidance and countermeasures, radio communications, and bomb destructiveness finally matched the early claims of air power enthusiasts. While men still flew the planes, computers took over bombing, navigation, and gunnery functions. Radar substituted for direct human vision; decoy and rocket-

127

General George C. Kenney, USAF, first commander of the Strategic Air Command, in the late 1940s when budgets were lean and equipment marginal. (U.S. Air Force Photo)

powered bombs supplemented free-fall gravity bombs; and guidance systems took over the pilot function at low altitudes. On the ground, automated display systems replaced the human croupiers of the old operations centers. When the ICBMs came in in the 1960s, fueling, check-out and guidance were fully automated. Only the act of firing remained in human hands.

As SAC became an electromechanical elite force, the men who comprised a human part of its total complex found themselves in an enclaved, almost monastic order, with a credo: "Peace is our profession." Long hours, high risk, and self-sacrifice were the regular lot of carefully selected and expensively trained crews. General Curtis LeMay, who commanded SAC during its period of greatest growth, fought the Air Force and Department of Defense bureaucracy for special privileges for SAC crewmen: payday twice a month, special housing for enlisted men and officers, sports-car racing on air base runways, free flying lessons for all ranks, and waivers on standard Air Force promotion regulations.[23] Privileged yet overworked, given the best equipment and the most demanding peacetime duty, first on call in emergencies

General Curtis E. LeMay, USAF, who symbolized the emergence of the Strategic Air Command as the mainstay of American nuclear diplomacy in the late 1950s. (U.S. Air Force Photo)

like the Cuban missile crisis, SAC drifted further from the "citizen army" model than any American peacetime military unit. The image of the Strategic Air Command as the backbone of American defense was strongest in the middle 1950s. President Eisenhower had ended the distasteful, frustrating Korean War, and Americans were sure there would be no more expensive, inconclusive ground campaigns in Asia. SAC was the symbol of Dulles's "massive retaliation" policy—"brinkmanship" to his foes—the line in the sand over which the Russians could step only at the risk of holocaust. While America held the thermonuclear edge, it worked. By 1960, however, the development of intercontinental ballistic missiles in the United States and USSR undercut the manned bomber. The U.S. Navy overcame its aversion to nuclear warfare as the *Polaris* system grafted nuclear technology onto an old naval elite corps, the submariners. At the same time, the SAC phenomenon illustrated a more general fact in American life.

The military had grown up. No longer short-budgeted and relegated to their own narrow society on remote military posts, soldiers, sailors and airmen were seeing the world. They were

129

A Convair RB 36D. This bomber version of the B-36 was the backbone of the Strategic Air Command in the early and mid-1950s. (U.S. Air Force Photo)

leading the way into the computer age. They had the best communications and transportation resources in the world. After 1941, American Presidents moved and communicated aided by military technology, and after 1945 lived a heartbeat away from the red phone, the link to nuclear war. That the presidency in the United States overshadowed Congress was related to the superiority of executive branch information systems and the fact that the President was the man who would have to decide to pull the thermonuclear trigger. The overbearing personal style of Lyndon Johnson and the Vietnamese debacle dramatized the fact that the executive had taken most of the marbles in terms of monitoring and control technology. The separation of powers had become a myth in view of the unprecedented marriage of presidential and military power. The executive branch of government by 1970 had thousands of computers, most of them in the Department of

130

Defense; Congress had two. And beyond that, the sophistication and education of the officer corps compared with that of congressmen was sometimes embarrassing.

Others were concerned that the American defense system had come to mirror the challenge of expansionist Communism. When Khrushchev dashed the world's hopes for an East-West accord in 1960 in Paris, Marshal Malinovsky at his elbow sat glowering, the unspeaking but powerful presence of the Red Army. There was a Russian strategic bomber force, an army ten times the size of that of the United States, and a submarine fleet ten times bigger than the Germans' at the beginning of World War II. Against this array, SAC was only one cybernated elite in America's Cold War arsenal. And then there were the CIA, NSA, ASA, CIC, ONI, OSI, FOI—a whole new bureaucracy of clandestine technology. Operations research begat computers; and computers and nuclear technology begat RAND; and RAND and thermonuclear technology begat SAC. For 150 years, the United States, long shielded by British sea power and foreign policy, had lacked an intelligence system.

After 1947, in a blend of apparent successes and failures, the Central Intelligence Agency, a civilian elite, had grown up. America got into the business of intelligence in an unprecedented way. The secret bureaucracies grew, but they were still upstaged by SAC, for the images of the massive, lumbering B-36 six-engined propeller-driven bomber, the sleek B-47, the massive B-52, the KC-135 tankers, the B-58 supersonic Hustler were closer to the American image of war: hardware and direct purpose. From 45,000 men in 1948, SAC grew to 200,000.[24] In the end it was technology, not politics, that eroded SAC's role and image. The manned-bomber enthusiasts, however, did not stop pushing for such a force when ICBMs came in in the 1960s. They argued the need of a deterrent independent of machine error, one

131

that could be used as a threat and recalled. As with the cavalry and battleships, there was a mystique and a social system involved like *Strategic Air Command* (1955), *Bombers B-52* (1958) and *A Gathering of Eagles* (1963). SAC's dedication to duty was symbolized by contrails in the sky before commercial jets came into regular use in the late 1950s and later by less popular sonic booms.

But in spite of the glamour, SAC was a burden, psychologically and financially, the modern American equivalent of the old Royal Navy. Even before Vietnam the incongruity of a honed, poised elite serving as the critical outer limit of American policy and military presence wore thin in the public consciousness. In Britain, far more vulnerable than the United States to thermonuclear annihilation, the Committee for Nuclear Disarmament marched and demonstrated against the "V" Bomber Force. American

The B-47, a difficult aircraft to fly, filled the gap between the propeller-driven B-36 and the B-52 in the late 1950s in the Strategic Air Command. With only an 800-mile range, it was designed to operate from bases around the rimlands of the Soviet Union. (U.S. Air Force Photo)

unease was more vague, but the bombers made people aware of the imminence of nuclear war in a way that the later missile systems did not. Perhaps it was only a case of visibility. The problem of bomber defense policy was aggravated by the strong ideological component in the SAC mission. Practicing to bomb Russian targets under dangerous conditions required commitment, dedication, and motivation. The prospect of going to war in a few minutes had to be accepted.

Maintaining enthusiasm was met by men in different ways. Some became students of Communism, and some drifted close to and over the line of political activism. Some SAC commanders became vocal in support of strong anti-Communist policy. General LeMay, then in retirement, ran for Vice-President in 1968 with George Wallace. In the Vietnam period, the military function itself became a target of criticism. Some professionals made the mistake of being drawn into the debate over the war as defenders of politically determined policy—thereby losing their professional detachment and immunity. The tendency of opponents in heated debate to suspect each other of the basest motives was merely enhanced.

To some revisionist historians pursuing a course of self-abasement, the Cold War was an invention of plutocratic conspirators and a budget-hungry military playing together in the "military-industrial complex," an updated version of the "merchants of death" concept of the 1920s and '30s. To paranoids on the other side, all critics of defense policy were Communist sympathizers. America's baptism of fire as a world power was painful indeed. SAC was a symptom of the age. But was it really an effective peacekeeper? Did it actually delay United States–Soviet accord? Was it a seedbed of neo-Fascism? Did its implicit role debase American values? Was it a true nemesis of Soviet aggression? Or merely a successful budget ploy by the Air Force?

133

The B-58 "Hustler" put the Strategic Air Command over the speed of sound in the late 1950s. (U.S. Air Force Photo)

Did its image and technology corrupt the executive branch over time?

We are too close to the events to see the answers clearly. But the pressure did exist for the U.S. Armed Services to indulge in marketing during the Cold War. The Army's "packaging" centered at various times on the airborne, the "Pentomic" concept, Special Forces, the air-mobile division, and strategic missile programs. The Navy went from nuclear and supercarriers to nuclear submarines. Throughout the 1950s, however, the older services were responding to Congress-as-appropriator-of-funds. And Congress did well by the Air Force's manned-bomber and ICBM programs. In the early 1960s the Kennedy administration supported the "spectrum of response" (conceived by Maxwell Taylor as a retreat from "massive retaliation"), expanding the Army on the eve of its "conventional" deployment in Vietnam. Unification

134

A Boeing B-52 of the Strategic Air Command takes on fuel from a jet-powered Boeing KC-135 Stratotanker. Under each wing, the B-52 carries a North American AGM-28 "Hound Dog" air-to-ground missile. (U.S. Air Force Photo)

of the services failed to alter the basic system in which they developed their own "product lines," resembling the large automobile companies in their use of markets research, styling changes, and the use of "public relations." During this period of wasteful and sometimes creative tension, one of the earliest cybernetic elites, the submariners, came to rival the Strategic Air Command's pre-eminence.

The submarine became the only weapons system thermonuclear delivery instrument to emerge outside of air force control. Submariners already were in elite status in the world's major navies by mid-century, a blend of specialized technology and rigid personnel selection.

135

They had not always been so careful in picking men, however. Only disaster forced a close review of the question, when accidental submarine losses in several navies in the 1920s and 1930s underlined the need for stability, compatibility, and proficiency in all crew members. Rigorous physical, psychological, and intelligence tests were developed to screen volunteers, and pay and food quality were raised. By 1941, for example, the U.S. Navy's submariners constituted such a rarefied corps d'elite that half the crew of the first U.S. submarine to go out on combat patrol after Pearl Harbor held commissions by 1945.[25]

The major submarine campaigns aimed at severing the lifelines of island empires. The Germans tried to strangle Britain in both world wars. The Americans assailed Japan's tender sea-lane arteries in the Second World War. The course of undersea war was shaped by the technology of attack and defense. In the First World War, the Germans stopped warning their victims and undertook sink-on-sight tactics when the British used disguised armed freighters—"Q ships." Allied refinements in depth charges and listening gear, along with convoying and primitive air escort, defeated the Germans.

In the Second World War, the margin of victory in the Battle of the Atlantic was perhaps narrower. The main burden of the battle fell upon the British and Canadian navies and their supporting science establishment. The techniques of "operations research" were used to develop new hunt-and-attack patterns and hardware like searchlights, radar, sonar, and depth charges. The Germans countered with detection decoys, special torpedoes, the "snorkel" for submerged battery charging, and the hydrogen peroxide submarine. Single-centimeter radar scored a bloody defeat on the U-boats in mid-1943. Light escort carriers, blimps, and extended long-range air patrols added to the pressure. The war ended before the Germans could deploy large numbers of their

Another elite corps were the men who served in submarines. Here a German U-boat is caught by a carrier aircraft on convoy escort. The year is 1943. (Official U.S. Navy Photo)

type XXI and XXII submarines designed to counter the Allied advantages. The Germans had lost over seventy-five percent of their U-boats and crews.

American losses were proportionally lower than the Germans', but high relative to losses among other U.S. ship classes. The Americans succeeded in sinking the entire Japanese merchant marine, plus some replacements. They were aided by resistance to change in the Imperial Navy. Japanese destroyer captains scorned the role of merchant vessel escort. They wanted to be the vanguard of the fleet in classic surface actions. Like all good bureaucracies, they preferred to do what they had always done, rather than adapt.

Like the Germans, the Americans found their torpedoes inadequate on early patrols, a problem soon overcome by research "boffins." By the end of the Second World War, aiming and firing torpedoes had become computerized, and submarine crews comprised a greater concentration of technical skills than those of other ships.

In the Cold War the expansion of the Russian conventional submarine fleet through the 1950s was met by a relatively casual NATO response. The 1960s, however, saw the development of the U.S. Polaris submarine. Antisubmarine warfare came back into vogue. The U.S. and Soviet fleets put nuclear-powered attack submarines—or antisubmarine submariners—into service. The British and French built their equivalents to Polaris. Submarine crews became an even more rarefied technological elite. Psychiatric screening was especially critical for men facing the possibility of being the first—or last—round in the chamber in a nuclear war.

Popular portrayals of the submariners' lot—such films as *Destination Tokyo* (1944), *Run Silent, Run Deep* (1958) and *Torpedo Run* (1959)—focused on interpersonal stress of the kind which psychiatric screening had attempted to eliminate. They minimized how electronics and machines had come to dominate the picture. Only *The Bedford Incident* (1966) and *The Enemy Below* (1958), dramatics aside, did justice in a popular vein to the technology in submarine and antisubmarine warfare.

Rear Admiral Hyman Rickover's battle with the U.S. Navy establishment to build a nuclear-powered submarine in the 1950s reflected a shift from the bluff-hearty, anti-intellectual professional officer—by no means extinct, but obsolescent—toward a new role of dedicated, efficiency-oriented "military manager." The nuclear-powered Polaris submarine was a change in basic U.S. Navy policy. For a decade after the Second World War, the supercarrier was the Navy's main budget magnet. Some sailors argued that the use of the Bomb was too horrible to contemplate, although carrier-based bombers did carry tactical nuclear weapons. An inversion in navy philosophy came with the blending of

138

Polaris with Oskar Morgenstern's "oceanic deterrent," a concept in which nuclear-powered submarines would become the sole nuclear "delivery system," thus reducing the Soviet tendency to aim missiles at air force bases and missile bases in the continental United States. The building of a fleet of nuclear-powered submarines weakened the carrier program. Vietnam brought the latter's withered state embarrassingly to the fore. But in the meantime, Polaris became a symbol of naval thermonuclear power rivaling Strategic Air Command's and RAF "V" Force's photogenic bombers.

But there was another dimension into which corps d'elite moved in the twentieth century. The idea of cybernetics came about as scientists and mathematicians became aware that me-

Submarines today have become a critical component in the arsenals of the superpowers. The Polaris and other programs have been the U.S. Navy's most successful budget magnet in the interservice rivalry for a thermonuclear delivery mission. Here the nuclear-powered fleet ballistic missile submarine USS Benjamin Franklin *surfaces off the island of Oahu.* (Official U.S. Navy Photo)

chanical and electrical systems could have intelligence and the ability to influence their environment. Fantasy writers seized on the possibilities of robots, leading to a new concept, that of the "cyborg," the cybernetic organism. In the case of SAC, submarines and Phantom worked *with* machines. But cyborgs *were* part machine, part man. Science-fiction stories described people who did not realize how much robot they were, or robots who did not realize they were not people, or people who forgot that robots were not people, and so forth. In the case of two elite forces that appeared in the twentieth century, however, the effect could actually be seen at work.

In the first case, that of the British Army's 79th Armoured Division in the Second World War, the men were representative of the parent force, but the machines were the elite component. In the second case, that of the frogman, men and life-supporting devices produced a bio-mechanical creature that could penetrate and move about in a previously hostile medium, the sea.

The 79th Armoured Division was perhaps the most unusual of twentieth-century ground forces. Its "Rube Goldberg"–"Heath Robinson" array of specialized fighting vehicles, such as "funnies," were highly successful but had surprisingly little impact later. In most accounts of the Normandy D-Day landings (June 6, 1944),[26] the 79th Armoured Division is either ignored or listed as a conventional British unit. One reason may be that it landed in segments along the whole British-Canadian beachhead rather than on a relatively narrow front as an integral unit. Yet Allied success at D-Day hinged on the 79th.

The 79th Armoured Division was formed in the spring of 1943 out of several specialized armor units and was commanded by Major General P. C. S. Hobart. After raising and training the 7th Armoured Division (the "Desert Rats") in North Africa, Hobart was retired on account of age in the summer of 1939 as a major general. After the war began, Hobart enlisted in the Home Guard. He was a corporal when Churchill intervened and recalled him to active service. An armor enthusiast in the 1920s and 1930s (who was also Montgomery's brother-in-law), Hobart has been depicted by defenders as a victim of peacetime military politics and the machinations of the cavalry clique.[27] But in his hostile contacts with experimental personnel and his martinet approach to troops, he displayed a crusty style that did him little good in the peacetime bureaucracy of smooth-talkers and flatterers.[28] And his task was not an easy one. It required coordination of many components. The 79th's equipment included amphibious "Buffalo" troop carriers; "Crocodile" flamethrower tanks; DD—"duplex drive"— amphibious tanks; "crab" tanks equipped with motor-driven flail chains to beat a path through minefields; bridging tanks; "Kangaroos"—armored personnel carriers made by removing turrets from obsolescent tanks; tanks for placing fascines; and AVREs (Assault Vehicle Royal Engineers), mounting a powerful short-range twelve-inch mortar, "the Flying Dustbin," for shattering casemates and bunkers and equipped with fittings for placing heavy explosive charges.[29] The unit was charged with developing the equipment as well as using it in combat.

The 79th served in different roles throughout the European campaign, in besieging the Channel ports, clearing the Scheldt, and crossing the Rhine. But its full inventory was vital at Normandy. In many accounts, the relative ease of British-Canadian lodgement compared with American frustration, particularly on Omaha Beach, has been explained as a result of bad luck. But there was prior Allied experience in landing against a hostile shore in northern Europe: the Dieppe "raid." There, in August 1942, a mainly Canadian reconnaissance-in-force was badly shattered. The disaster nevertheless taught many lessons. Reports were subsequently distributed to U.S. and British commanders to aid

141

in planning for D-Day. But the U.S. tactical planners ignored or rejected many of these dearly won experiences.[30]

During the invasion buildup several senior officers, including Eisenhower, Bradley, and Montgomery, visited Hobart's unit. Eisenhower, the Supreme Commander, was enthusiastic and asked for "everything you can give us." Montgomery, who commanded Allied forces on the Continent until D-Day + 90, listed specific requirements. Bradley, the American tactical commander, said he would discuss needs with his staff.[31] They used only "DD" tanks from the 79th's array of "funnies." The price was paid by U.S. engineers, who strove unshielded, at great cost, to breach the German defenses. In a further departure from orthodoxy, the 79th was not a regular division per se but was custom-built and flexible, able to perform various operations components under the control of GHQ. The 79th's CDL units, for example—made up of high-intensity searchlight tanks—never saw action for a variety of reasons, mostly bureaucratic. The division was disbanded after World War II, its only vestige being a Specialized Armour Development Establishment at Woodbridge in Suffolk. The concept of the "funnies" was virtually forgotten, even though General Eisenhower in his report on Normandy said of the 79th: "It is doubtful if the assault force could have established themselves without the use of these weapons."[32]

In a general sense, looking beyond its successes, the 79th was also a break with the armor tradition, for it did not use tanks to shield men as much as it used men to guide highly specialized machines. Perhaps because the next obvious step was eliminating human crews, once innovative and iconoclastic, specialized armor à la Hobart did not live on after 1945. Armor had arrived and developed its own orthodoxy. The armies of the world continued to field a limited range of standard tanks, assault guns, and personnel carriers.

142

An unusual technological elite force, assembled especially for the Normandy invasion, was the 79th Armoured Division, comprising bulldozing tanks, minesweeping tanks and other vehicles designed to clear the beach of obstacles Allied intelligence warned were waiting. Pictured here is a Bobbin or carpet-laying tank, designed to bridge soft sand. It is based on the Churchill tank chassis, and mounted in its turret is a 105mm howitzer. (Photo courtesy Imperial War Museum)

In contrast, the frogmen evolved in the direction of a man-machine system slowly. But the concept survived. Clandestine swimmers, after all, are hardly new to warfare. But a new technology of underwater breathing developed in the twentieth century, which allowed free movement, revolutionized the tactics of underwater warfare. As with the use of high-speed motorboats, the Italians wrote the first and the most spectacular chapters of the story. In the First World War, Italian junior officers destroyed three major Austrian warships in the Adriatic with motorboats and torpedoes aimed by swimmers. These efforts were not institutionalized. On the eve of the Second World War a new generation of junior Italian naval officers revitalized the program. In 1935 the Italian invasion of Ethiopia led to threats of war between Italy and Britain in the Mediterranean. In that crisis, the 10th MAS *(Motoscafi Antisubmergilio)* was formed. The title of a motorboat flotilla was a cover for an experimental force which developed tactics and technology for attacking harbors with small submarines and with swimmers using underwater breathing gear. The ebbing of war fears put the 10th on the shelf until 1939 and the coming of war in Europe. When Italy entered the war in 1940, the 10th saw action along the North African coast and around Algiers. Its most spectacular success was the sinking of two British battleships at their mooring in Alexandria harbor[33] and of Allied cargo ships at Gibraltar using magnetized "limpet" mines. After the 1943 armistice, MAS veterans passed on much information to the British and formed a parachutist-frogman unit which fought for the Allies in the northern Adriatic, a concept which lives on in contemporary form in the U.S. Navy's SEAL program.[34]

British frogman operations began as a response to the Italian challenge, using veteran salvage divers at first. One- and two-man torpedoes undertook special operations, including an ineffective sortie against the German battleship *Tirpitz* in Norway. The Germans undertook a similar program of their own and used frogmen against Allied shipping at Anzio and against bridges on the Rhine with limited success. The American underwater warfare effort was mainly directed at preinvasion beach reconnaissance and demolition. In spite of a relatively slow start and primitive equipment, the U.S. Navy's Underwater Demolition Teams played an effective part in many landings in Europe and the Pacific with relatively light casualties. The first UDT teams went into the beach from motorboats, wearing only goggles, trunks, and rubber fins, carrying special tablets and styluses to record obstacle data. Shore bombardments distracted the enemy, since they often operated in daylight within sight of friends and enemies.[35]

The use of SCUBA (self-contained underwater breathing apparatus which recycled oxygen for chemical purification to avoid telltale bubbles) was more prevalent in Italian, British, and German forces in World War II than in the U.S. Navy. But even that gear limited the free diving depth to about 30 feet.

A Cold War *cause célèbre* brought advances in underwater technology to the public eye in 1956 when Commander Edward Crabb, a pioneer British frogman who had won the Victoria Cross in the Second World War, disappeared. Some speculated that he was killed or abducted while checking the hull of the Soviet cruiser *Ordzhonikidze* which was visiting Britain. Rumors circulated that he had defected, since he had been passed over for promotion. Official British comment was vague.

The increased interest in oceanic exploration was stimulated by a French naval officer, Jacques Cousteau, who had developed the diluter-demand compressed air "aqualung" and increased the depth of mobile diving tenfold. The search for undersea resources, the development of "oceanic deterrent," and expansion in world submarine fleets stimulated research in these areas, leading to experiments with membranes to allow direct breathing of sea water, special gas mixes for deep diving, and the use of porpoises

145

A U.S. Navy SEAL team on a raid southwest of Saigon in January 1968. Note the variety of weapons carried. The SEALs are versatile all-purpose combat troops for special missions, and in practice have no more relation to naval warfare than the British Special Air Service had to aerial warfare. Their training (sea, air and land) is similar to, though less comprehensive than, that lavished on the Army's Special Forces. (Official U.S. Navy Photo)

for combat patrols in the Mekong Delta during the Vietnam War. Although space exploration seemed to be demilitarized, the competition for knowledge and control of the undersea world gave promise of confining conflict.

Looking at the cybernetic elites as a species, one can see that the accent on youth and exotic expertise increased steadily, and that the old social communities of the military world were shattered. So was the ability of commanders and civilian policy shapers to understand or even to keep up with developments in so many areas. The growing exchange between theoretical science and the professional officers bypassed political monitoring. The new elites were unique not only by military standards but also in terms of science and technology. The old military elites had been kept in line by close budget review in Anglo-Saxon countries at least. Now the new systems required long-range planning and budgeting which made annual review impossible. The growing dominance and immunity of the new elite organizations were obvious to audiences that laughed at *Dr. Strangelove.* The fear of coup d'état produced such dark versions of the future as *Failsafe* and *Seven Days in May.* But there was no tangible threat. A coup would not be needed in a nation where power was abdicated piece by piece to the military system, or where its technology made the government by contrast corrupt, ineffectual or obsolescent. Even the left wing was happy when the Army toppled Senator Joseph McCarthy in the 1954 hearings. The most serious damage done by the cybernetic elites was their unintentional undermining of the citizen-soldier concept and universal military training, for they gave the illusion that victory could be bought by substituting gadgets for human beings. The illusion was often broken, but the idea of robot legions lived on, obscuring the fact that the effect of employing mercenaries is the same on their employer whether they be men or machines.

A specialized elite: one of the U.S. Navy's Underwater Demolition Teams in action. These teams were first organized to neutralize Japanese beach and harbor defenses during the reconquest of the Pacific islands in World War II. There were, however, earlier "frog-men," notably some Italian units in World War I. (Official U.S. Navy Photo)

VII

IMAGE AND IDEOLOGY:
ILLUSIONS OF ELITISM

C ORPS d'elite in the twentieth century were more than just militarily useful. They could serve a publicity and propaganda function, and they were also good show business. Images of paratroopers, mountain troops, and commandos dramatized war in an age when the modern battlefield was desolate much of the time. Gun cameras caught a glimpse of aerial action, jerky, brief, and fragmented. Photographs of troops moving in the open, or in direct combat, were scarce. Only one picture was taken in combat from a U.S. Navy surface ship of a Japanese surface ship in World War II. In this vacuum of imagery, scenes of training and mock combat substituted for real action. At the same time, propaganda was growing as a tool of war. The U.S. Office of War Information, the British Ministry of Information, the Nazi Propaganda Ministry, Tass, Sovfoto, the propaganda agencies of democracies and totalitarian countries shared a common value: news and images were to be controlled and shaped. Purpose varied more

148

than technique. The British and French did a fairly skillful job of softening up Americans for World War I and II. The Nazis and Fascists used propaganda so regularly that many stopped believing what came out of those systems, even though they were often less devious than their opponents'. In this world of manipulated ideas, half truths and lies, elite forces of various nations came to represent values as well as functions. The Greek Sacred Company, *Hierchos Lochos,* composed completely of officers of the Greek Army, embodied Royalism among Greek forces fighting with the British in the Mediterranean during 1941-44.

In some cases, image surpassed reality. The greatest paradox was the Communist use of elite forces. Fascists were social elitists, so their use of special units was natural enough, but the Communists were egalitarians, levelers. Yet in Russia, China, and Spain the leaders of the masses created elite forces to compensate for the deficiencies of the masses. This embarrassing contradiction was evident at the birth of Soviet Communism. In the Bolshevik revolution of 1917, the backbone of Lenin's attack on the Provisional Government were *ipso facto* elites, the sailors of the Baltic Fleet, the Kronstadt naval garrison, and the Lettish Rifles. Communist success in October–November 1917 is often portrayed as a victory of workers and Red Guards, but the Bolsheviks depended on formal military units in their move for power. The revolt of Baltic sailors against the Reds in 1921, which changed the nature of Soviet Communism, has overshadowed those earlier events.

The "storming" of the Winter Palace, which brought down the Kerensky government in 1917, appears in Soviet mythology akin to the Bastille, or to the Alcazar, stormed by Red Guards and the workers. But sailors secured vital points in Petrograd and paralyzed the Provisional government's technical network. Later, the Bolshevik victory was made final in a series of small actions when

149

Kerensky made a last try for power, and again the sailors played a vital role. Kerensky made his escape disguised as a sailor.

After the Russian Civil War of 1918-22 a debate between elitists and populists erupted in the Red Army command. This led to a compromise between Mikhail Frunze, who wanted a general militia, and Tukhachevsky, who sought a highly trained professional force. The result was a mix: 560,000 regulars and a forty-three-division territorial militia. Frunze was closer to Friedrich Engels's thoughts on a "nation in arms," while Tukhachevsky was more in line with Marx and Lenin in proposing a revolutionary elite. Eventually, Stalin moved to professionalize the Red Army, a goal he shared with Leon Trotsky and Tukhachevsky, both of whom perished opposing him.[1]

For the first twenty years, the Red Army had rejected the forms of the old system, e.g., epaulets, saluting, formal feudal ceremonies, modes of address, and unilateral command systems. But the Finnish fiasco of 1939 and the massive defeats of 1941-42 brought a change as the struggle against the Nazis became the Great Fatherland War—a patriotic defense of Russian soil, not of the Communist state. Party generals were eased out. The posters, films and periodicals of the era reflect how Stalin saw nationalism as a better glue than socialist ideals.[2] Saluting, officer privileges, and tight discipline returned, and the commissars faded away.

In early 1943, the 1st Guards Army was created, including many of the parachute veterans who survived the purges of the late 1930s and the early battles of the Hitler war. Units designated as "Guards" received extra pay, special privileges—and extra artillery support.[3] Eventually eighty divisions[4] received such titles, suggesting a general sop rather than true elite status. When the morale of the Soviet Army's occupation forces declined in 1946, the Guards designation, being phased out, was retained as a permanent title for some units.[5] By the end of the Second World War gaudy uniforms, decorations, shoulder boards, saluting, and tough discipline were firmly entrenched.

Not surprisingly, the Chinese Communists also compromised with the problem, for the leadership of the Chinese Communist forces and the government developed from an army built by surviving leaders of the 6,000-mile Long March of 1934-35. In the late 1930s, after the Long March, they adopted military discipline and organization not unlike Cromwell's New Model Army, and, following the Russian model, awarded elite status after the test of battle. The highest honor in the People's Land Army, the title of "Iron and Steel Brigade," was given to division commanders before an impending battle. Divisional sub-units were then "allowed" to compete for the coveted designation.[6] The Chinese pragmatically adopted Western military systems from the German, Russian, and American advisers—and their opponents, the Japanese. Communist Chinese forces became a hybrid system with strong emphasis on morale overcoming superior military technology—and military elitism and unit spirit as a source of cohesion and initiative.

In the 1960s they established "model" or demonstration units in the air force[7] and "Four Excellence Companies" in the army.[8] Until the late 1950s the Chinese Communist militia was an "elite . . . a carefully picked minority."[9] This and the fact that the 1949 victory came after "units of the Kuomintang Army defected to lay the basis for the Red Army"[10] again reflect the paradox. The collision of ideology with power produced a drift to militarism, not utopianism.

In the army of the German Democratic Republic, a similar tension between classlessness and the need for "special" forces appeared. The Soviets built the East German Army—with skillful propaganda maneuvering—around Nazi veterans. The Ministry of State Security formed a small army of its own, reduced in the early 1960s to a Guard Regiment of 4,000. In uniforms and behavior it resembled the caricature of such forces in television spy series. After the building of the Berlin Wall in 1961, the Frontier Troops became the most visible elite force. They received special military *and* political training and an official elite designation in an attempt to counter the alluring blandishments of decadent capitalism so close at hand. But frequent desertions of officers continued.[11]

All these cases were consistent with Lenin's emphasis on elitism. The betrayal of popular revolutions in France, China, and Russia was by those who used military organization as the keystone of power.[12] The bloody suppression of the Kronstadt revolt of 1921 with massed artillery and gas saw the triumph of the centralized model of state Communism. But in the end, it ground the suppressors under as well as the suppressed; only one Soviet leader of that period avoided a violent death.[13] And in China in the late 1960s the battle among the military, the bureaucracy, and the ideologues was won by the army. A principal advantage of these "Communist Warlords" in this struggle was effective communications technology, linking together "an interregional, interfield army 'party' balance-of-power system."[14] Once again, a resort to force shattered the illusion of Communist populism. Militarism has remained a strong Soviet and Chinese propaganda theme. In the half century after the Bolshevik revolution, the reliance of Communism on traditional military values and organization was more of an ebb tide than a wave of the future.

The most powerful nemesis of the Communists so far was another ideological elite force, the *Waffen* SS, the *Schutzstaffeln* of the Nazis. Inasmuch as the SS was an organization built as a symbol, in the end image outlived effectiveness, and the SS wandered far from its original form and purpose. The SS was organized in April 1925 as a small bodyguard to protect Hitler from

Motorcycle units of the Waffen SS. *The machines are BMW RW75s, and the machine gun mounted on the side car is the standard Wehrmacht MG34. Battalions of motorcycle troops, organized into four or five companies of eighteen motorcycles each, were often attached to SS Panzer and Panzer Grenadier divisions for reconnaissance and cavalry screening, or to act as shock troops in supporting infantry assault operations.* (United Press International Photo)

attack by political opponents—as well as from the *Sturmab-teilung,* Roehm's brownshirted SA, the Nazi street army. The SA was openly revolutionary, but SS members were silent on political matters. Heinrich Himmler, its founder, and later head of the "SS state," saw the SA as "infantry of the line" and the SS as "the imperial Guard of the New Germany."[15]

The SS grew slowly, receiving infusions of *Freikorps* veterans in the late 1920s—elite by their own definition at least—and followed after 1929 by middle-class professionals thrown out of work by the Depression. The rampant homosexuality of the SA led to publicized trials during the tense election battles of the early 1930s. Himmler eliminated such members from the SS, setting the stage for the "Night of the Long Knives" in 1934 when Hitler used the SS to buy the Army's favor by purging unsavory elements of the SA—and some innocent bystanders.

In keeping with Hitler's view of the need for "individual formations" to fight "titanic battles" in a "revolution of the Spirit,"[16] the SS expanded in the late 1930s into larger formations, the *Verfügungstruppe,* designated as *Waffen* (armed) SS in 1940. The totemry, trappings, and ceremonies were shaped by Himmler in accord with his romanticist impulses. In this period aristocrats appeared in the higher ranks and in the SS cavalry, but most left after the Roehm putsch in 1934,[17] and Nazi "eugenic" selection standards were imposed.[18] The SS cadet schools, various fumbling experiments with autarchy and selective breeding, the Burgundy scheme for creating a National Socialist enclave, and Himmler's bizarre ceremonies and costuming, suggested to some—including Hitler—a religious order patterned after the Jesuits.[19]

Peacetime SS selection standards had been based on appearance, physical condition, and bearing. But the war changed the SS into an umbrella under which Orientals, Muslims, and Slavs joined Nordic elements in a vast multinational army. Of the 600,-

000-man, thirty-eight-division *Waffen* SS, only the 18th SS Panzer Division (Horst Wessel) maintained a level of performance which made it recognizable as a really elite field force.[20] "Racial purity" was well watered down as the SS encompassed groups once defined as subhuman by the Nazis. Excesses of the Dirlewanger Division—actually a penal unit—and the slaughter of unarmed prisoners and civilians by some junior officers further stained the SS image. Toward the end of the war, the all-German *Waffen* SS divisions (seventeen of the thirty-eight) closely resembled Wehrmacht units in performance.[21]

When the *Waffen* SS began to play an expanded role on the Eastern Front in 1942-43, its relatively high losses became common knowledge, and recruiting drives met resistance in Germany. Some Wehrmacht officers suggested that the political selection of officers caused such squandering of valuable human resources.[22] Rommel discouraged his son from joining the *Waffen* SS in 1943 on the grounds that Himmler would be too unsavory a commanding officer.[23]

Nor were those the only frustrations that faced Himmler in trying to build his empire within the Reich. The SS failed to recruit the peasantry, particularly the Catholics, in any appreciable numbers in spite of an elaborate *Blut und Boden* (blood and soil) mythology. Even the praetorian status of the *crème de la crème* crumbled in the last days of the war when the Führer's SS bodyguard, the *Leibstandarte Adolf Hitler,* remained alive in the ruins.

But many were tough; many did die. The *Waffen* SS served well as a "fire brigade," an emergency reserve in the last days of the Third Reich. Their deeds of daring and honor and their sinister glamour captured the imagination of succeeding generations. Like Napoleon's colorful but defeated. legions, they became popular models not only for military buffs and collectors, but also for

155

admiring former foes who pointed to distinctions between the *Einsatzkommando* and the relatively clean record of combat units. Former SS officers in West Germany often found their credentials an aid in job-seeking, and *Waffen* SS gained veterans' pensions from the Bonn government in 1961.[24] Men of the *Waffen* SS, moreover, continued to fight in the field in Indo-China and Algeria as many former SS, in the spirit of Ernst Jünger, sought "The Fight for Its Own Sake" in the French Foreign Legion.

Postmortems of the record of the SS at Nürnberg, and historians' analyses, revealed a full spectrum of behavior. Adherence to the laws of war by some *Waffen* SS units and individual acts of kindness blur into nihilistic danger-seeking *à la* the French and Spanish Foreign Legions and then on to spontaneous massacres and finally into the darkest shades of cruelty and clinical sadism. Similarly, comparing *Waffen* SS with *Allgemeine* SS units charged with extermination and terror roles—the *Sicherheitdienst, Totenkopfverbande,* and *Einsatzkommando*—reveals a range from chivalry to the sinister and depraved. The scenes of SS training and indoctrination recounted by early SS members[25] are a synthesis of Bosch, Wagner, Disney, and Goya. It is not really the hard evidence of the SS that constitutes a problem, as much as the spread of its image through fiction, film, and folklore as tough, select, dedicated bands who stood off the world in spite of the foot-dragging of an apathetic or disloyal Wehrmacht. The SS upstaged the French Foreign Legion as a popular image of military elitism. Its fame has outlasted many of the forces that brought it down, and it continues to gain respectability as the generation who felt the direct horror of Nazism fades away. It may well have a delayed impact, a deferred inheritance of the Thousand Year Reich. The tree was cut, but the seeds are still on the wind.

The use of military elite units as potential symbols was not confined to totalitarian countries. The increase of literacy through

A U.S. Navy PT boat patrolling off New Guinea in July 1943. A good deal of Japanese naval tonnage was lost to the torpedoes of the PT Boat Service. Since that time, PT boats have been dramatically used at the hands of the North Koreans, the North Vietnamese, the Egyptians and the Israelis. (Official U.S. Navy Photo)

general education and the penny press parallels the growth of liberal democracy and the general franchise. In the twentieth century, technologies of propaganda and intelligence were turned more and more into the business of shaping and manipulating public attitudes in countries where leaders were elected. The most dramatic example was the use of journalism, mass advertising and propaganda by the Nazis. But there were others.

In 1960, massive amounts of money spent on media brought to the forefront of American politics John F. Kennedy, a young man who traded heavily on military affairs, and corps d'elite in particular. Even though he had a weak attendance record in the U.S. Senate, and even though he and his family had supported Senator Joseph McCarthy, he was presented as a dynamic liberal

157

and a vigorous young crusader. His poor physical condition was masked by a public image of vigorous health. And in the forefront of his campaign was the symbol of his war record, the PT boat. He had been a junior naval officer in command of a motor torpedo boat which had been cut in half by a Japanese cruiser at night. The accident was blown up by exaggeration and repetition into a major incident of the Pacific war. At the same time, it was strongly implied that PT boats were a vital part of the war effort and a corps d'elite. Silver PT boat lapel pins were given to key campaign workers. In this case at least, image far transcended reality. The Kennedy ploy was to build on the PT boat image puffed up in the early days of World War II when mosquito boats were picturesque and a favorite of action-starved publics in the United States and Britain. Unheard of before 1940, by 1960 the motor torpedo boat became such a symbol of dash and power that the Kennedys used it as a unique symbol in their campaign for the presidency. Those who worked in their organization were later awarded a sterling silver lapel pin, a profile of John Kennedy's PT 109. Like fighter planes (no matter how badly they performed in actual service), the mosquito boats were photogenic, suggesting individual verve and heroism, and drew lyrical support from journalists and editors. The evacuation of General MacArthur and his family from the Philippines in a PT boat was dramatized in a book and later a film starring John Wayne and Robert Montgomery.[26] Less well publicized was the fact that the Italians, who pioneered the use of fast torpedo boats early in the century, were most successful with this medium over time.[27]

Torpedo boats were used in the areas of critical confluence around Europe, in the North Sea, the Mediterranean, and in constricted waters in the Pacific around the Solomons, New Guinea, and the Philippines. While the British faced effective and numerous counterforces in the Channel and Mediterranean, the

Japanese did not raise a force equivalent to the U.S. PT boat squadrons.

In terms of losses, impact, and personnel selection the mosquito fleets were not really elites. In spite of journalistic images[28] of David and Goliath battles, motor torpedo boats spent much of their time on patrol, escort, and shuttling duties. Crews were not especially picked, and the operational statistics[29] indicate their combat effectiveness. Useful in narrow waters for shuttling and patrolling, they were not a major arm of the fighting fleets outside the area of the English Channel and the Mediterranean.

	British	German	Italian	U.S.
Boats commissioned	1,383	570	144	561
Boats lost	223	209	74	79
Warships sunk	8	40	? (2+)	0
Merchant ships sunk	140	99	?	(mostly barges and lighters)

The sinking of the Israeli destroyer *Eilat* in 1967 by a missile fired from a torpedo boat fifty miles away made some experts think the torpedo boat might be on the ascendancy. If such predictions prove correct, it would not be the first time that a weapons system became effective only long after confident claims to its decisiveness were made.

Students of "popular culture" can find much valuable material in comparing such images of military effectiveness with the facts unearthed later. The contemporary versions of the realities of trench fighting were so far off the mark that one correspondent, Philip Gibbs, wrote a chronicle of his involvement with journalistic lying and censorship in World War I. Many books that came out during World War II on ordnance, weapons, and aircraft now appear ludicrous. The power of the media to distort kept pace with its ability to reveal. Military policymakers, then, have two directions to look when considering the creation of elite units: forward, to apparent military purpose; and over their shoulder at unpredictable public images that may emerge. For they are creating not just organizations, but symbols as well. The early reliance of John Kennedy on the PT boat image has been obscured by his later dependency on another force, the Special Forces, more truly an elite unit, and one which generated many unpredictable side-effects in terms of image and subsequent impact on policy. The illusion of effectiveness and predictability is an omnipresent one, and constitutes the more dangerous cutting surface of a two-edged sword.

VIII

THE ULTIMATE PASSION:
THE KAMIKAZE ATTACK CORPS

FOR three months in 1945, the island of Okinawa and the sea around it blossomed with fire and smoke. Army and Marine Corps divisions were locked in battle against a 100,000-man Japanese Army. At sea, the fleet that came to provide the troops with support and supplies was assailed from the air. As expected, because they were within range of Japan, there were regular bombing and strafing attacks. But there was something new. The sky became filled with planes, dozens of planes, bent on crashing into the American and British ships—wave after wave. The fleet filled the skies with its antiaircraft fire. Fifty-caliber Brownings, 20-millimeter Oerlikons, 40-millimeter Bofors, five-inch guns, drowning out speech and thought in a steady roar, like a great blowtorch, until sea and sky were splattered with a curtain of flashes and splashes and puffs until the horizon disappeared like a great animated Jackson Pollock painting. . . . And still they came, on through the curtain. It did no good just to hit them once or twice. They would come on, sometimes a great ball of flame, relentlessly. They had to be destroyed, the pilot killed. . . . No, that was not enough; for some, rigid in death, still came on. The planes had to be demolished in midair.

The supply pipeline was alerted and reserve ammunition stocks, whole mountains of steel and copper and high explosives, were consumed to keep the umbrella of steel and fire up against the rain of fanatic, nightmarish death. The sword of *bushido* was again cutting deep. A new word, a synonym for destructive suicide, was introduced into the language of man: *kamikaze*. . . .

"The bowmen and knights of fourteenth-century England looked out on the Thames one morning and saw an armored steamship squadron lying at anchor. . . ." What if feudal, preliterate, agricultural England had been dragged headlong into the

The ultimate elite: Japanese kamikaze pilots assemble to take up their aircraft and die. This photo was taken in 1945. (Courtesy U.S. Naval Institute)

machine age? What if the superstitious cults and feudal ideas had been synthesized with industrial technology in a single generation? Add isolation, a cultural reverence for suicide, a divine monarch, and take away the Magna Carta and the yeoman tradition, and the model becomes more and more like that of Japan. The forcible opening of Nippon to Western trade by Admiral Perry in 1859 produced a strange and at first fascinating blend of old and new, East and West. The British advised on naval affairs, the Germans on the army; and the Japanese learned so quickly that they were soon a rival to the major Western imperial powers in the Far East by 1895.

But changes came too quickly and too unevenly. Bushido, a new warrior code, the synthesis of technology and the old samurai warrior spirit, produced an unexpected result, a violent caricature of Western imperialism. Japan seized Formosa from China in 1895, and in the Russo-Japanese War of 1904-05 her navy won first-class power status, and she began to move out from her home islands aggressively. An alliance with Britain gained her more possessions in the First World War. A cancellation of that alliance in 1922 by Britain further stirred Japan's ambition to become master in Asia.

In 1931 she moved to seize Manchuria, the industrial heart of China. The pacifist Western democracies and the impotent League of Nations talked much and did little. By 1937, the liberal-imperialist junior officer factions in the Japanese Army terrorized the government, and a series of campaigns in China began that finally led to collision with the United States in 1941. American firmness and trade embargoes and the rise of the Japanese Army war faction to control over the government brought on the tensions that led to Pearl Harbor. When the Japanese fleet air arm finished its work on December 7, 1941, the back of the U.S. Pacific Fleet was broken. The smoke pall over the lush green hills of

Hawaii seemed to mark the grave of American power in the Pacific.

The Japanese then turned to American, British, Dutch, and Australian positions in the Pacific, central and south, and Southeast Asia, all of which had to be neutralized for Japan to replace her quickly dwindling oil stocks from the Dutch East Indies. No one, not even the Japanese, expected such success. Hong Kong fell, the western islands in the western Pacific fell, Singapore fell, then Java, Sumatra, Thailand, Timor, and the Celebes. The Philippines held out longest, and then succumbed in May 1942. The Japanese carrier task forces smashed Allied naval power in the South Seas and ranged from Ceylon to Hawaii and from New Guinea to the Aleutians with impunity. When they pushed the British out of Burma in the summer of '42, the humiliated Western powers, faced with Hitler's victories in North Africa and Russia and the steady sawing of Nazi submarines at the Atlantic lifeline, talked in near-panic of "too little and too late." Except for the bombing of Tokyo by a dramatic sortie—the Doolittle raid—there was no sign of an ebb in Japanese fortunes.

The Sons of Heaven had won against the Chinese in 1895 on Formosa. They had scattered the Russians in 1904-05. They had overrun German possessions in China in 1914-15. They had pushed the Chinese out of Manchuria and away from the coast. Now they had broken the power of the West in Asia. It was heady stuff for a nation that believed itself descended from the gods, and it produced a case of indigestion. A debate raged in the Japanese high command over the next course of action. Attack? Withdraw and consolidate? The result was Midway, when luck suddenly changed and five aircraft carriers, the heart of the Japanese fleet, slid beneath the waves on June 3-4, 1942.

Much hard fighting was ahead in the Pacific war, but America had wrenched its back away from the wall in those few dramatic hours and proceeded to rain the most devastating series of blows on an opponent in the history of warfare. Steadily, American industry and Allied manpower caught up with the Japanese lead. By late 1943, Japan faced Allied countermoves in Burma, the central Pacific, the Aleutians, New Guinea, and in the Solomon Islands. Overextended, the Japanese now found that the space they had taken was their worst enemy. Yet the sea lanes to the oil of Indonesia had to be kept open. Sixty divisions were tied up in China. A full field army stood on guard aganst possible Russian moves in Manchuria. They had spread too much, too far. But there was no way out.

In January 1943 at Casablanca, President Roosevelt and Prime Minister Churchill announced their demand for unconditional surrender. The stakes were survival. Neither the Japanese nor the Nazis would accept such a fate. And so it went on, until Berlin and Tokyo's hollow ruins echoed to the tread of their most feared enemy, and millions more died.

If the West had industry and machines, the Japanese had courage and a contempt for comfort and life itself that astonished their opponents. They charged in waves—the so-called *banzai* attacks—and fell by the thousands, raked and shredded by the wall of steel and explosives laid down by the Americans, the British, and the Australians on island after island, battle after battle. By 1944, the superiority of American carrier pilots over the Japanese was obvious. The last big carrier plane battle was called the "Marianas turkey-shoot" by the U.S. victors.

As the pressure mounted, the Japanese leadership delved deep into the darker recesses of their tradition for something to stave off the relentless tightening of the vise. The Japanese suicide corps, the kamikazes, the ultimate in twentieth-century military aberrations, was their response. By the autumn of 1944 the Japanese air forces had lost most of their experienced pilots. American fliers

An example of what the kamikaze attacks could do: the aircraft carrier USS Franklin burns out of control in 1945. (Official U.S. Navy Photo)

were now coming to combat with two to three times the flying experience of their opponents, in ever increasing numbers from a growing fleet of aircraft carriers, with faster and more heavily armored planes. An American army had landed in the Philippines following a series of sea battles that had gutted the Japanese Navy. The U.S. Navy Marine Corps were more than two-thirds up their ladder of advance through the central Pacific that led to the home islands. The bulk of the Japanese Army in China faced exhausting attrition, and their forces in Burma were beginning to starve after the collapse of an offensive against India. The Japanese merchant marine had been virtually sunk by American submarines; the B-29 raids on Japanese cities from Chinese bases promised the holocaust to come. Tens of thousands of Japanese troops languished on Pacific islands, bypassed by the dazzling advances of American amphibian strategy.

The suicide theme has run strong in Japanese cultural tradition.[1] The world knew the stereotypes—romantic tales of young lovers hurling themselves into volcanos and ceremonial suicide, hara-kiri, the samurai's grisly act of atonement for dishonor. The Americans, British, and Australians had been amazed by the repeated futile banzai charges and the tenacious defense of positions to the literal last man. It was not a complete surprise to the Americans when the Japanese went the final step and formed a unit whose purpose was suicide, the kamikazes. There had been many individual suicide attacks by Japanese soldiers and aircraft in situations where death was an alternative to capture, or for a beneficial sacrifice, such as crashing a crippled plane into a hangar or diving into an American torpedo about to hit a carrier. Those who committed these pre-kamikaze acts were mainly from the aristocracy and the lower middle class, the primary supporters of the Japanese samurai tradition. The formal organization of the "Special Attack Corps" attracted a new kind of volunteer—

167

religious fanatics, the more generally patriotic, and yet others affected by what the Japanese viewed as cold logic.[2] "Kamikaze" (Divine Wind) came from the name given a great typhoon that destroyed an invasion fleet of the great Khan bound on the conquest of Japan.

The limited and sporadic kamikaze attacks in the Philippines in 1944 did not seem notably different to the Americans from the usual spontaneous suicide attacks. In the summer of 1945, when the Americans landed on Okinawa within air range of the Japanese home islands, the kamikazes emerged as a "weapons system" in the most bizarre sense. At Okinawa, only American technological advantage and Japanese deficiency denied a victory to the grisly new force. The U.S. arsenal included sophisticated radar and proximity fuses which detonated near-misses of antiaircraft guns by radio signals reflected from the target aircraft. Outlying radar picket ships allowed U.S. planes to intercept and gun crews to prepare for Japanese sallies. At the same time the Japanese aircraft were lighter than their opponents'. Even with high-explosive bombs attached to them, they could not detonate beneath the waterline of the American ships; nor could they, as naval shells could, penetrate the armor plate of heavier ships before exploding. British carriers hit by kamikazes were relatively unscathed since they had armored decks—which American carriers lacked.[3]

The Japanese tried to compensate for inadequate explosive power and penetration with such innovations as the bomber-transported *Oka,* a rocket-powered bomb with stubby wings. It was unsuccessful. Since the kamikazes were a compensation for the lack of pilot skills, the use of a high-speed vehicle like the *Oka* made little sense. Its chronic failure led the Americans to dub it the *Baka*—or foolish—bomb.

In spite of all the technical shortcomings, kamikazes did fearsome damage, as reflected in the following statistics:

U.S. Naval Losses at Okinawa [4]

4,900 killed; 4,800 wounded

Ships sunk:
    9 destroyers
    2 minesweepers
    2 transports
    1 subchaser

Heavily damaged:
    20 major fleet units (including 1 carrier permanently
    retired from service)
    74 destroyers
    78 smaller units

Eighty percent of the losses were inflicted by the kamikazes.[5] The ferocity of the sea war off Okinawa surpassed that on land. The U.S. Navy lost more than U.S. ground forces in that largest battle of the Pacific war. But the U.S. fleet stayed and Okinawa fell. It was not a total defeat for the Japanese, however. In Washington, plans were in final process for Operations *Olympic* and *Coronet,* a two-stage conquest of Japan by direct naval and ground assault. Projected casualties were one million men. U.S. and British forces were being transferred from the European theater. The Japanese were nowhere near finished. It was estimated that they had hidden 5,000 to 6,000 planes in caves in the home islands, and that in a major invasion a fiercer kamikaze assault could be expected against more tightly concentrated and vulnerable targets.

The Allied leaders, therefore, faced a potential bloodbath in the late summer of 1945. They did not know that kamikaze morale was declining as the fact of imminent defeat percolated into the lower ranks of the Japanese services.[6] They did know that the Japanese were mustering *kaiten*—explosive-laden motorboats, and *kairya*—submarines—to attack an invasion fleet. The 4,364 kamikazes gave better than they got numerically, even though in some cases they were not volunteers.[7] They created new respect

169

for Japan's defensive power, even though her merchant marine and navy were virtually sunk, her Chinese and Kwantung and Burmese armies were stranded, and her cities were being snuffed out one by one with fire-bomb raids. Certainly, the kamikazes created the climate in which the atomic bomb was dropped on Hiroshima and Nagasaki. But when the dust settled, the Japanese avoided the fate of the Kaiser, the Austrian emperor, and the Czars, King Boris and King Peter, and dethroned monarchs of the twentieth century. Their emperor, Hirohito, kept his throne, and Japan received kind treatment from her captors. The kamikazes cast a long shadow.

It is tempting to see the Divine Wind as a peculiarly Japanese phenomenon, a living example of the excesses of bushido. But the theme of death is common in the chronicles of elite forces. The motto of the SS, "Our honor is our loyalty," suggested death before dishonor. The French and Spanish Foreign Legions have a strong self-destructive ethic, and the Swiss Guards of the Bourbons proved their dedication by the final sacrifice. True, Japanese feudalism remained unchanged longer than any other and was grafted onto modern industrial imperialism with dire results. But were the kamikazes a bizarre flower in the Japanese garden, or a variant on a more universal species? After all, the self-immolation of the defenders of Masada, the Alamo, Chapultapec, and the farmhouse at Camerone are seen as sublime by Israelis, Texans, Mexicans, and Frenchmen, respectively. In all these cases, the defenders died to the last man. There are enough such examples to suggest that the kamikazes were merely an extreme case. Whether it is a case of a "death wish" is a matter for academic or psychiatric speculation. But membership in all elite corps carries with it a latent role as a bridegroom of death. The kamikazes were the ultimate.

170

IX

THE SELECTION-DESTRUCTION CYCLE

THE decision to form an elite gets into the area of statistics. If you have a thousand people chosen randomly, all of these won't be enthusiastic about doing a job, whatever it is. Now, if you call for volunteers for a dangerous mission, how many of all these will volunteer? Let's say a third. Of those who volunteer, how many will pass the medical exams and the training that you set up to get them ready? Let's say half. How many will be left at the end? Half of a third, or one-sixth. Then, you send your special group off to do their bit for the parent organization, king, country, God, Yale, or whatever. The hazardous mission proves hazardous, all right. Your special unit suffers fifty percent casualties.

Now, you must sit down with your graph paper and your scratch pad and tote up the butcher's bill. Did they succeed in their mission? If so, was it worth the cost? Of course, you say? Their bold deeds frightened the enemy, forced diversion of reserves, caught them off balance, made them look flatfooted, and cheered the hearts of the special group's comrades and countrymen. All right, you feel it was worthwhile. But what do you know? What are your data? How did you calculate the cost? Did it solve all your problems? Aren't you still going to have to take what's left of the general organization and use it in battle against the enemy? Did the special mission really make further fighting unnecessary? Or did it just get the enemy's back up?

But what about costs? We can take a closer look at that. What did you do by selecting all those eager fellows and putting them into a tight situation where losses were ten times what the general organization usually loses in battle? What if those chaps had been trained as the leaders of your general organization instead? For one thing, they would have been spread around, so that their overall loss rate would have been lower. But now you don't have them anymore . . . Oh, you have half? Are you going to disband that glorious special unit as a reward for all its special achievements? Are you going to lose all that *esprit* and show the general organization that the reward for doing a good job is to be dropped from the rolls? That's leadership?

So the special unit will remain intact. The best one-sixth of your whole group is gone, one way or another. So you ask for volunteers to fill the leadership roles in the general organization. Are these going to be up to the standard of the lost one-sixth? In combat, air forces have found that about five percent of the fighter pilots shoot down about fifty percent of the enemy planes. The rest shoot down the remainder of enemy kills. Of course everyone can't be a star. They haven't the capacity, right? So what have you done by forming your special group? I'll let you answer that.

Oh, yes, and while you're doing your cost accounting, what's the effect going to be if you do this kind of thing on a nationwide scale instead of with just a thousand men? What will the effect be if you skim off the best men in war and get them killed off at a high rate? The boys who survive are going to go home and have kids, right? If they can. What if there's a genetic dimension to all this? Well, then, either you've found a good way of getting rid of dangerous, violent, overeager people—or you've found a good way to strip your organization of leadership over the very long run. Maybe both. And for organization, maybe you should substitute "nation." Who knows? Have a nice time with your graph paper.

Criteria for selecting members of elite forces were fairly simple until the late nineteenth century. Men were recruited on the basis of height, condition of teeth, general appearance, "soundness of limb" and the like. There was no medical screening worthy of the name until science began to amass data which made selection objective. Eyesight tests, for example, did not appear in the British Army until the 1870s. But in the twentieth century, volunteers for corps d'elite were chosen by increasingly precise medical and psychological standards. As psychological and skill tests were refined in industry, there was cross-pollination between the worlds of commerce and the military. The rise of aviation made effective physical and mental analysis vital. Pilots needed good eyes, high resistance to vertigo, good circulation, sound nerves and fast reflexes. Selection was only the first phase, for means had to be worked out to sideline fliers whose condition created a safety problem.

Science was far less effective in weighing leadership ability. Attempts to screen on a psychiatric basis were not successful in the U.S. or British forces in the Second World War. After 1945, studies focused more and more on the physiological aspects of stress and anxiety.[1] More general psychiatric screening was singularly unsuccessful in predicting combat effectiveness even well into the Vietnamese conflict.[2] And, as with lie-detector tests, there was much aversion to using such tests in democratic countries, perhaps even more so if they showed evidence of effectiveness. The British Army's program, adapted from German experiences, finally became the subject of parliamentary debate. Liberal elements and conservative soldiers shared a dislike of the program. The former saw psychiatric evaluation as an invasion of privacy and an unwarranted extension of state power. The latter preferred their own intuitive and, as they saw it, proven methods of selection. The traditionally elite Guards regiments ignored them.[3]

Science not only aided the sharpening of selection tools, but was also invoked to design training systems which could further sift those who were already volunteers. Yet, at the end of this process of distillation, the final products were committed to battle conditions where concentrated technologies of war produced disproportionately high casualties. The losses of airborne forces, Commandos, Rangers, and U.S. and British bomber crews were extremely high compared with general force levels, which had dropped dramatically from the First World War. The low level of refinement in selection during 1914-18 had been caricatured by George Grosz in his cartoon of a German Army medical board certifying a rotting corpse as fit for duty. In the late days of the Second World War, standards again dropped dramatically in Germany. "Stomach battalions" were formed—units of men with chronic ulcers requiring special diets. Old men and children were drafted into the *Volksturm,* an improvised second-line reserve force. In a grimmer sense, the new techniques of warfare in industrialized countries followed a parallel track to these overall selection refinements which produced a bloody paradox. Better means of picking the best men was paced by building new tools of war, which increased the slaughter among the products of the refined selection process.

174

Thus three-quarters of the British 1st Airborne Division were lost at Arnhem in ten days.[4] One-third of the German parachute forces died on Crete in two days. Of the 33,302 members of the British Glider Pilot Regiment committed to action, 551 were killed and 1,301 wounded.[5] The high loss of airborne forces in Sicily led Allied planners to expect a similar pattern on D-Day, since, for example, fifty British gliders had crashed into the sea off Sicily and the Allied fleet had accidentally shot down twenty-five planeloads of paratroops over Salerno harbor. This led the SHAEF planners to write off seventy percent of the glider troops and fifty percent of the paratroops in Operation Overlord, the invasion of Normandy.[6] This led to a behind-the-scenes drama, a debate between Eisenhower and British Air Marshal Leigh Mallory on the eve of the landings. Leigh Mallory tried to stop the airborne phase of the operation, and Eisenhower overrode the Air Marshal's objections; but "Ike" was seen in tears when leaving the preinvasion staging camp of the 101st Airborne Division. Fortunately, Leigh Mallory's pessimism was misplaced. Casualties were less than ten percent. But the odds caught up at Arnhem. And in spite of their low relative combat exposure, loss rates of the airborne divisions in Europe equaled those of the ground divisions.

The fate of the U.S. Marine Corps parachute units was similar—even though they never made a combat jump. "Paramarines" were used as conventional ground troops and as raiders early in the Pacific war. In the late summer and autumn of 1942, on Gavutu-Tanambogo, they suffered fifty percent losses a few days after commitment to combat and were the first Marine unit withdrawn in the Solomons campaign. The pattern was repeated when the badly mauled Marine Provisional Parachute Battalion was pulled off "Hellzapoppin' Ridge" on Bougainville. Even at the time, it was recognized that the slaughter was due to their having

Members of the American 10th Mountain Division and other units, in 1944, being instructed by Italian Alpine troops on the use of taut equipment lines to remove casualties. (U.S. Army Photograph)

been committed to sustained combat with inappropriate weapons.[7]

The Germans overlooked their own heavy losses at the outset of airborne warfare, owing to the frantic pace of events in 1940. In the first-wave attack against The Hague, forty-two percent of the officers, twenty-eight percent of the enlisted men, and ninety percent of the transport aircraft were lost.[8]

In another dimension, the U.S. 10th Mountain Division, the most elite U.S. division in the twentieth century in terms of intelligence scores, fitness and training, suffered five times more casualties preparing for combat than any other American division in the Second World War. When they were committed to action in Italy against German mountain troops, the 10th suffered the heaviest casualties relative to time in combat of any U.S. division in the Italian campaign.[9]

The selection-destruction cycle, however, was most apparent in the Second World War in the air and in submarine warfare. Both systems had once seemed to offer a way to wage war on the cheap by avoiding costly pitched battles. But defensive tactics changed, requiring alterations in the offense. Each subsequent refinement generated a reaction by the opposition, which crushed the dream of bargain victories. Only the armored forces, where emphasis was placed on protective machines as opposed to selection of personnel, experienced a marked reduction in combat casualties. But there were always exceptions, and attempts to beat the system were often frustrated. A recent RAND study shows, for example, that the relationship of the activity ratio and the initial force ratio in the Battle of Britain matched that of ninety-two historic land battles, suggesting that "the same pattern of dependency" may be "a general characteristic of all armed combat."[10]

Such cruel data destroys the illusion of air war or submarines as uniquely modern. Rather, the image comes to mind of ponderous medieval knights lifted onto horses with pulleys, supported

177

by armorers, squires and serfs. Aviators and submariners became the focus of advanced metalworking technology, as had the chevaliers. The "knights of the air" received special status, identity and privilege, while their vast support organizations and the achievements of the standard forces often remained anonymous. Aces were touted, but no one knew or cared which infantryman or tank gunner scored the most "kills." Sometimes the parallel with medievalism was not subtle at all. The men of the U.S. 8th Air Force wore personal battle armor in the later phase of the European air war. And the devastation of aerial bombing recalls the *chevauchie,* the medieval practice of ravaging the land to deny an enemy supplies and discourage his followers. What eluded the dreamers of an easy way of war was the fact that technology only amplifies human impulse, be it good or ill.

It was, nevertheless, by publicizing air power as a new guarantor of victory that air forces in many countries received primary access to quality manpower. The U.S. Army Air Forces were able to maintain the highest selection standards in the Army, even at the lowest ranks, well into World War II.[11] In 1943, over forty percent of draftees in the top two of five intelligence categories were assigned to the USAAF as enlisted men. In officer selection, performance tests and physical examinations further refined the sample, so thoroughly that the Air Forces' Aviation Cadet Classification Battery has been judged as "probably the most successful attempt yet made to measure a special kind of aptitude by measuring a complex combination of traits." For example, seventy-eight percent of those tested in the lowest of nine categories failed in actual training and field service, against only five percent of those in the highest category.[12] It should be kept in mind that all of those tested were volunteers. Even when critics from the other branches of the U.S. Army pointed out that outstanding men were being wasted in marginal ground roles in the

USAAF, Air Force medical officers felt that they never reached preferred standards.[13]

In a more bizarre case, the U.S. Army's personnel system virtually broke down because of errors in planning in late 1943 and early 1944 when projections of infantry and other ground force replacements made early in the war fell short of actual demand. In the frenzied search for manpower, the Army turned on its *crème de la crème,* those highest scorers on the Army General Classification Test, sent to college for special training. These men, along with expensively trained, carefully selected air cadets and technicians from service branches, were committed to combat as infantry fillers.[14] Considering that replacements suffer high casualties under ordinary circumstances, this affair had the flavor of bureaucratic insanity, *à la* Franz Kafka and Lewis Carroll, a perversion of the selection process.

Rigorous selection was not a peculiarly American practice. For example, Japanese pilot training before the Second World War included an initial wrestling competition to weed out the weaker candidates. In 1937, the Japanese enlisted-pilot program chose seventy candidates from 1,500 applicants. After continued ordeals of hazardous gymnastics, underwater and speed swimming and calisthenics, along with ground school and flight training, thirty-five survived to receive pilot status. Japanese standards were lowered, however, in the course of the Second World War, while American standards were raised.[15]

What happened to all these carefully cultivated flowers? The Japanese Naval Air Arm, so carefully nurtured, was virtually cut down by 1944. In the U. S. Army Air Forces, 36,515 officers and 49,514 enlisted men were lost during the Second World War from all causes. About 25,000 of these were lost in accidents.[16] The USAAF lost 52,172 of the Army's 234,874 battle deaths.[17] More Air Corps officers were killed in action than all other Army

179

branches combined, and Air Corps enlisted losses exceeded all branches except the Infantry. In all, the Air Forces lost about 27.7 percent of Army deaths from all causes.[18]

British air losses were higher in combat and lower in training. Bomber Command lost 47,268 men on operations and 55,888 in all—seven percent of Britain's wartime manpower. And, as was the case with the Americans, these men were all selected, trained volunteers.[19] Even though the Soviets rejected the concept of strategic air power, they too suffered heavy losses, with over 70,-000 admitted aircraft losses in the Great Fatherland War.[20] While this casualty figure only suggests actual air crew losses, the Soviets saw much of their air force chewed up on the ground in the first phase of Operation Barbarossa, the German invasion of Russia in 1941. Subsequent training, moreover, was under high stress, hurried under primitive conditions.

Air war also led to the deaths of many who might have survived if wounded on the ground. Air crew wounds on planes that returned safely often occurred at the most distant point of the flight. First aid was complicated by the crew's preoccupation with flight and fighting off attack. Temporary lack of oxygen or protective clothing could aggravate shock and produce severe frostbite or outright freezing. Administration of plasma was difficult at best. The reading of vital signs by distracted amateurs in a crippled, yawing, vibrating, drafty aircraft with a damaged intercom at temperatures of 20° to 30° below zero at night was hard enough. Ministration of medical aid would be impossible beyond morphine injections. And pilot error or his wounding or death in a bomber could mean a dozen deaths, not merely one.

A similar loss pattern appeared in submarine warfare. Although U.S. sub crew casualties were considerably above the Navy combat average,[21] the Germans lost ninety percent of their combat U-boat crews—38,000 men. Submarine crew selection was rigorous in all navies. The Germans measured strength of motivation by the use of electric shocks, filming the subjects and analyzing their reactions. Interestingly, U.S. and British military psychologists focused on intelligence and skills, whereas the Germans were interested in willpower.[22]

Even though the selection-destruction cycle became more noticeable in the Second World War, some dimensions of it appeared earlier. In view of the reputation of the Australians as first-class infantry during 1914-18, for example, it is interesting to note that forty percent of all Australian men ages eighteen to forty-five who were eligible for military service went into the AIF in the First World War. Of these 417,000 men 330,000 went overseas; 60,000 were killed and 160,000 wounded. Of all Australian males of military eligibility, six percent were wounded.[23]

As one looks back at all these cold statistics, a configuration emerges. It would be convenient to explain such recurrent slaughter with some concept like, say, Gaston Bouthoul's theory of infanticide[24] in which he hypothesized a hatred for youth by the aged, resolved by inventing war to kill off the threat. Such a view would require acceptance of a rather bold premise. But in looking at the gap between what was expected from elite forces and what they did, it is easy to argue that Clausewitz's observations on the role of change in war and Lenin's view of war as unmasker of governmental weakness merge. Emphasis in modern political systems on "progress" and immediate "results" and the endless competition of promises has led to an ever wider gap between being and seeming. When wars have come, as von Seeckt suggested, they usually have come as failures of policy. The grim course of these conflicts, lurching out of control of their creators, might suggest to a poet a monster at large.

Yet the corps d'elite further underscores the chronic failure of modern governments to control the flow and the result of warfare.

181

The illusion of close central direction, supported by apologists and memoirists and assisted in hindsight by historians and journalists, lives on. In contrast to this myth of rationally waged war, the gravestones of the elite forces loom up. They form a pattern which shows not order and control but waste and futility. They are also guideposts for those who hope to avoid the swamp, especially in view of the influence that the corps d'elite image has had on political and military leaders. If the illusion that these units present of a responsive, effective instrument were not at hand, would policymakers have been so ready to resort to war? The general experiences of twentieth-century military elite forces demonstrate little of glory or hope. They indicate, rather, that the sacrifice of youth and resources has too often been in vain. As with the Light Brigade, the achievements of the brave and eager have been magnificent but rarely worth the cost. Only the genetic and eugenic science of the future may be able to tell what the cost really was.[25]

Linked with the selection-destruction cycle is another distortion, the gap between the role for which elite forces were created and the role that the actual conditions of war forced them into by the time they were raised and trained. The cases already examined of the 1st Special Service Brigade, the Commandos and Rangers, the airborne, Merrill's Marauders and the Chindits and the bomber fleets come easily to mind. There were other cases as well.

The U.S. Army Tank Destroyer Corps[26] was formed in 1941 as an attempt to develop antitank techniques in the face of what seemed to be an overwhelming German armored superiority. Human courage was to make up for mechanical disadvantage, and the TDC was given full branch status to reinforce the importance of its mission. It was involved in a tank-killer role early in the war, in North Africa, but as Allied tank production surpassed that of

182

the Germans, the tank destroyers were used more and more as assault guns and as indirect artillery support. They drifted so far from their original mission that they were disbanded in 1946.

A similar pattern was seen on a much broader scale in the case of mountain troops[27] used in large numbers in high country in the First World War and to a lesser extent in the Second. By 1943, the bulk of the German Army's mountain forces were serving as infantry on the Russian front, and their training standards had been eroded. Others were assigned to duty in Italy and Yugoslavia. The British mountain division, the 78th, saw action only on the plains of northwest Europe, and the U.S. 10th Mountain Division, highly trained and carefully picked, saw only a few months' action in the Apennines. After 1945, the use of such units shrank even further.

All of these cases show gravity at work, a pressure toward general function, not specialization. It is, of course, a cliché of ecology that crisis favors generalists and stability favors specialists. This, naturally, refers to the habits of animal species in their habitats. But it also parallels Lanchester's equation for victory,[28] which decrees that the best way to win wars is to mass one's forces and drive straight for the enemy center. If war is crisis—and it seems rather obvious that it is—and if Lanchester is right, then corps d'elite, unless they mesh with a special technical function, are merely ways of sidestepping the question. One must, at this point, look away from military function and examine instead the political dimension. One might ask if Western leadership outside the neutral countries has not gotten into the habit of killing off the flower of its youth rather than risking the political feedback that might come from casualties suffered by the generally raised forces. In Vietnam, even relatively few losses toppled the Johnson government. In a cynical vein, in view of that experience, it seems that the best balance of military forces in a society with antimilitary values and an elected government would be the use of elite forces and of conscripts and/or recruits from politically impotent elements of the population.

X

COUNTERPOISE TO MASS-MAN:
THE FUTURE OF ELITISM

ELITE forces are blooming still in the last third of the twentieth century; species are many and hues intense. Rangers, paratroops, and marines serve as the sword and buckler of the South Vietnamese in the wake of the American withdrawal. A rejuvenated Soviet marine force, wearing white berets, seems to be a hybrid of the quite different models of the British Royal Marine Commandos and the U.S. Marine Corps.[1] The "Derry massacre" in 1972 made British paratroopers conspicuous once again as constabulary. The American Special Forces, expanded widely in the 1960s, seem to be fading away even before enough facts have surfaced to treat them as history rather than legend. The continual toying by the U.S. Army with a revitalized Ranger concept underlines the persistence of an American self-image of elite soldier-frontiersman. Different varieties of that model used in Vietnam—the Recondo teams, Long Range Reconnaissance Patrols, CIA mercenaries, SEALS and Phoenix forces—remain

The culmination of military skill, specialized training, taxpayers' burden, and elitism: part of a Special Forces A Team in action. Better known to the public as Green Berets, the Army's Special Forces were trained to fight guerrilla warfare behind enemy lines, but were frequently used—as in Vietnam—in a counterguerrilla role. (U.S. Army Photograph)

shadowy, part rumor, part legend and part fact. Internationally, frogmen are a flourishing and vital species in the opening up of undersea colonies, perhaps the first generation of true "cyborgs."

In view of such continued vitality, it seems safe to predict that elite forces will thrive on, even when, from a rational standpoint, they constitute a waste of resources. The portrayal of their experiences belongs as much to poetry as to science or even history at a time when the dull, shabby vistas of industrialized society have led men to try to brighten their landscape. In that sense corps d'elite are a kind of military psychedelia, an irrational response to uniformity or overload, a larger-than-life product of popular culture. Thus traditionally pacifistic Jews have come to identify with the triumphs of Israeli paratroops and fighter pilots. The image of martial triumph is, after all, powerful. If man is still a paleolithic hunter at heart, these echoes of the hunting-warrior band are crucial, and the tension between such groups and civilization is not merely academically interesting. Military elite units, therefore, constitute social indicators which no politician or soldier can ignore. As with a powerful storage battery, care must be taken to tap the energy of such forces gently and slowly and to recognize that such reservoirs of power are not inexhaustible.

Such questions as whether privates in the South Vietnamese corps d'elite should be sergeants and lieutenants in the larger forces are the same kinds of problems faced by the Germans with their storm troops in 1918, by "Bomber" Harris with the Path Finders in 1942, by the Americans in the Second World War when appraising the need for airborne and special army divisions, and by the Germans with the *Waffen* SS. This dilemma will confront others in the future who create or control such organizations. Interestingly, the profile of military elite forces is low in the neutralist nations of Sweden and Switzerland, but high in Spain and

Portugal.[2] While cultural and historical differences are obvious, the latter countries maintain an imperial posture, casting a role for elite forces as a response to incipient rebellion.

In view of the scale of militarization in the United States itself since 1940, the American experience with such elitism is particularly interesting. The spectacle of the least militarized large nation emerging as a principal military power has been startling. The irony is increased by the fact that a main theme of American military history from 1775 to the Second World War was the debate between supporters of a popular militia and those who preferred an elite, professional force.[3] The question seemed relatively academic in peacetime from 1789 to the late 1930s, when the U.S. Army was barely large enough to provide auxiliary police services. Not surprisingly, some American officers envied their European counterparts in the late nineteenth and early twentieth centuries when conscription and new technology made mass armies—and high ranks—the norm on the continent.[4] The problem had been incipient, however, since the 1840s, when American production pulled up alongside that of Britain.[5] The chronic failure of the militia in the War of 1812 and the uneven behavior of volunteers in the Mexican War set regulars against the citizen-soldier concept, while Congress from the Revolution on supported the populist model of military amateurism. There were anti-military traditions as well, older than the Revolution. The first British regulars did not come to America until after the Glorious Revolution of 1688. The burden of defense in the colonies fell on locally raised militia. In the first half century of the Republic, the Army's mounted troops had to be called dragoons because of Congress's objection to aristocratic "cavalry."

While there were socially elite units in the militia, in war elite status came after the fact, as it did to Ringgold's artillery in Mexico and to the Iron Brigade in the Civil War. The American antimilitary sentiment was reinforced in the late nineteenth century by European immigrants fleeing European conscription—and even Civil War veterans opposed a professional general staff. The argument over creating one followed much tragicomic fumbling in the Spanish-American War, in which only Spanish ineptitude averted disaster. In 1904, Elihu Root, Secretary of War, published the report written by Emory Upton,[6] a regular officer in the 1880s, after a tour of European armies. Upton concluded that American disasters in the Civil War came from lack of a general staff system. The battle raged until 1907, when President Taft intervened and brought the General Staff into secure status. This was followed over the years by a blending of militia and regulars in wartime, and in peacetime as well, in terms of organization, equipment, and standards, under the control of the Regular Army, although tensions between reserve interest groups and the active forces are still evident.[7]

In the Cold War, American antimilitarism sank to a low level. The proliferation of corps d'elite in peacetime was a new thing in the U.S. services and paralleled the decline of the U.S. Congress's power relative to the presidency. And during the same period a mythology of America as a nation skilled in arms blossomed, beginning on the eve of World War II with such films as *Sergeant York* and *Northwest Passage*. After 1945, documentary and fictional television series took up the theme: for example, *Crusade in Europe, Air Power, Crusade in the Pacific, Victory at Sea, Combat,* and *Twelve O'Clock High.*

There were relatively few countercurrents: the problem of prisoner-of-war behavior in Korea;[8] the collapse of universal military training in 1951; the fiasco of the Reserve Forces Act of 1955; and the noisy reluctance of reservists called to active duty in the Berlin Crisis of 1961. S. L. A. Marshall's findings on the low levels of combativeness among the infantry in World War II[9] were unpublicized, as were unflattering comments by former adversaries.[10] Strident attacks on the British by Omar Bradley, Elliott Roosevelt and Ralph Ingersoll[11] and Cold War hostility toward Russia and China reinforced the American sense of having played the key role in both world wars. Except for brief embarrassment in Korea, the nuclear bomb kept the question of the fighting quality of conventional American forces out of sight until Vietnam put a strain on a broad manpower base. As long as the burden of the Cold War fell on virtually voluntary draftees and the men of the corps d'elite within the larger services, traditional American apathy toward military affairs persisted, even though the hostility had ebbed away.[12] And even for those forgotten men, the standard of living was relatively high, as it had been for U.S. forces in both world wars compared with other nations.[13]

In the early phase of the Vietnam War, 1964-67, Special Forces, airborne and the Marines bore the brunt. Dissension in the United States rose with the increasing use of conscripts. The American advantage in firepower and technology in the world wars and Korea[14] was blunted in Vietnam where nonlinear warfare denied the massing of targets that the American system was designed to destroy. The debate over the war reflected the difficulty, if not impossibility, of a liberal democracy fighting a distant and limited war with conscript forces in large numbers. Less generally noted was the inability of the elite forces to provide an effective brushfire war response beyond a short and shallow level.[15] Interestingly, since the end of the Second World War, the Russians and Chinese have maintained large universally conscripted and elite components in their armed forces.

The Western nations, unlike the Communists, became less and less inclined to consider any dimension of military policy unless looked at in terms of defense or a rationale of preserving the peace. As they became less physically and mentally prepared to face war, they became voyeurs of violence. In the United States, the decline in liberal-populist demands for closer civil control of the armed forces from 1945 to 1960 led to a reconcentration of military power. By abdicating military defense to corps d'elite, Western leaders confronted an old quandary, perceived by Joseph Shumpeter: if organizations are created which gain status and reward from threats, those organizations will seek out and dramatize threats.[16]

This change of course in warfare was noted long before "wars of national liberation" received Soviet official blessing in 1956. In 1950, F. O. Miksche prophesied new proletarian forms of war which would negate bourgeois military technology.[17] In predicting a blurring of the boundary between war and gangsterism, Miksche's prophecy was reminiscent of the sociologist Gaetano Mosca's view that demands to abolish armies came from violent radicals who hoped to dominate society themselves with their relatively small forces. Mosca asked: "When a war has ended on a large scale, will it not be revived on a small scale in quarrels between families, classes and villages?"[18] A glance at the thirty-odd conflicts simmering in the early 1970s gives Miksche and Mosca full marks for anticipating a redefinition of military victory. Continued growth of corps d'elite can be expected where short-range emergencies are the rule.

Elitism is a function of organization, since organizations tend to become stratified as they grow and as they persist. Should, however, this tendency be encouraged? Hills and valleys make beautiful scenery, but the more splendid the spectacle, the more difficult the terrain and the greater the danger. If a high standard of living, general education and urbanization weaken people's will to maintain military forces, then corps d'elite should multiply at a time when the power of machines relative to men in such forces is also growing. The visions of holocaust haunted World War I veterans like Henri Barbusse and J. F. C. Fuller.[19] Yet it may be that they were reacting to the actual zenith of military destruction. Nuclear weapons and the mechanization of war have caused a fragmentation of war. Now the revulsion of the slaughter wrought by heavy technology since 1939 has brought visions of the ritualization of war to the point where no people would be involved. Battles would be waged by expensive but bloodless robots.[20] There have been, of course, many dreams in the realm of military theory and design. The robotic olympiad may come about, but in the meantime elites will fill in the gap.

There is one other caution label that should be put on the medicine bottle marked "elitism." Elite forces are virtually encapsulated delinquency in many instances. Like guerrilla bands or street gangs, corps d'elite adopt strange customs and habits and costumes unrelated to any specific purpose. Just as guerrilla activities blend into the world of the gangster, some elite forces have assumed the posture of virtual hoodlums. Delinquency among youth has long been recognized as a function of "peer group membership," which when it "takes the form of a contraculture, gains prestige and permanency . . ."[21] The kindling of such youthful tribalism by a civilization should not be masked by high motive. Regression is regression, no matter how thick the disguise. Since almost every elite force of the twentieth century has cut close to or across the laws of war at some point, the following description could apply equally well to youth gangs or elite units: "They have an affinity for the romantic role of outlaw, which is perhaps the only status in which they feel they can stand out as individuals."[22]

Where the use of youthful zeal and aggressiveness has been a cosmetic for the weakness of modern military systems, it was a crime against those called to sacrifice themselves; insomuch as such images and legends will serve as models for the future, it was

192

a crime against posterity. The symbol of elite forces, like a strong antidote, has been used for positive results in spite of potentially dangerous and unpredictable side-effects. The ultimate significance in the proliferation of corps d'elite in the twentieth century, aside from those with a specific technological function, is as symptoms of stress in institutions reaching desperately for nostrums in crisis or moral bankruptcy.

References

CHAPTER I
ELITE UNITS: A HUNGER FOR HEROES

1. Robert McQuie, "Military History and Mathematical Analysis," *Military Review,* May 1970, 8-17.
2. See Alex Inkeles and Daniel J. Levinson, "National Character: The Study of Modal Personality and Sociocultural Systems," *The Handbook of Social Psychology,* ed. Gardner Lindzey and Elliot Aronson, 2nd ed. (Reading, Mass., 1969), pp. 457-458; and Stephen Rosen, "Reading Our Culture Through Books," *Wall Street Journal,* September 16, 1971, 14. In this vein, it has been suggested that literature immediately after a war reflects horror and that gradually revulsion is suppressed as memories age. As an increasing number of a society's members have not had firsthand experience with violence, themes shift from "horror dominant to glory dominant," tied to the human life cycle of twenty to thirty years. See Frank H. Denton and Warren Phillips, "Some Patterns in the History of Violence," RAND Corporation Paper P-3609, June 1967, 20. It is not, therefore, surprising to find General Chennault relating that in his youth he "pored over the history books in my grandfather Lee's library, reading about the Peloponnesian and Punic Wars . . . enthralled by the charging elephants, armored warriors and burning ships . . ." See Claire L. Chennault, *Way of a Fighter: Memoirs of Claire Lee Chennault* (New York, 1949), p. 5. This echoes a similar observation: "From the *Iliad* and the *Odyssey* to . . . present-day movies and television, the hero is obliged to pass a series of tests of his martial valor. . . . The message of heroic literature is clear: to

194

win the embrace of his beloved, the hero must accept the embrace of war," by Santiago Genovese, *Is Peace Inevitable? Aggression, Evolution and Human Destiny* (New York, 1970), p. 23.

3. See essays on these views in *An Introduction to the History of Sociology,* ed. Harry Elmer Barnes (Chicago, 1965). More recent sociological views of elites are less admiring. Andrzejewski's theory of "Praetorianism" and Wolfgang and Ferracuti's "subculture of violence" show little optimism about the final fruits of elitist evolutionary development; Lang, Janowitz, Bouthoul, and Moskos have been more carefully critical. See Stanislaw Andrzejewski, *Military Organization and Society* (London, 1954), pp.105-107; and Marvin E. Wolfgang and Franco Ferracuti, *The Subculture of Violence* (London, 1967), pp. 158-161.

4. S. L. A. Marshall, *Men Against Fire* (New York, 1966), pp. 10, 56-57.

5. *Soviet Military Strategy,* ed. V. D. Sokolovski, trans. Herbert S. Dinerstein, Leon Goure, and Thomas W. Wolfe (Englewood Cliffs, N.J., 1963), p. 334.

CHAPTER II
CHILDREN OF FRUSTRATION: CORPS D'ELITE IN THE FIRST WORLD WAR

1. C. R. M. F. Cruttwell, *A History of the Great War* (Oxford, 1936), pp. 431 ff.

2. Lynn Montross, *War Through the Ages* (New York, 1946), pp. 741-743.

3. Hans Ernest Fried, *The Guilt of the German Army* (New York, 1942), pp. 164-168.

4. For detailed discussions of the tactics involved, see Barrie Pitt, *1918: The Last Act* (New York, 1962), pp. 43, 46; B. H. Liddell Hart, *The Real War: 1914-18* (Boston, 1930), pp. 391-392; Office of the Chief of Military History, *American Military History, 1609-1953* (Washington, D.C., 1959), p. 343.

5. Heinz Höhne, *The Order of the Death's Head* (New York, 1970), pp. 444-447.

6. A. Rossi, *The Rise of Italian Fascism,* trans. Peter and Dorothy Watt (New York, 1966), pp. 229-231.

7. Christopher Seton-Watson, *Italy from Liberalism to Fascism, 1870-1925* (London, 1967), p. 498.

8. Herman Finer, *Mussolini's Italy* (Hamden, Conn., 1964), p. 114; Roy MacGregor-Hasti, *The Day of the Lion: The Life and Death of Fascist Italy, 1922-45* (London, 1963), p. 69.

9. F. L. Carsten, *The Rise of Fascism* (Berkeley, 1967), pp. 49-70.

10. Seton-Watson, *Liberalism to Fascism,* p. 571.

11. Ninetta Jucker, *Italy* (New York, 1970), p. 78.
12. Marjorie Barnard, *A History of Australia* (Sydney, 1967), pp. 477-478.
13. R. M. Younger, *Australia and the Australians* (New York, 1970), pp. 623 ff; also see C. E. W. Bean, *Anzac to Amiens* (Sydney, 1961); L. L. Robson, *The First AIF: A Study of Its Recruitment* (Melbourne, 1970); and Winston S. Churchill, *The Second World War,* Vol. II-VI (Boston, 1948-53).

CHAPTER III
HITLER'S ONLY VICTORY: COLLISION OF ELITES IN SPAIN

1. George Mills, *Franco* (New York, 1967), p. 111.
2. Jose Larios, *Combat Over Spain* (New York, 1966), p. 34.
3. Luis Bolin, *Spain: The Vital Years* (New York, 1967), pp. 86-89.
4. Brian Crozier, *Franco: A Biographical History* (London, 1967), pp. 54-72.
5. For a statistical analysis of Asturias casualties compared with the Irish Troubles, see Mills, *Franco,* p. 195.
6. Hugh Thomas, *The Spanish Civil War* (New York, 1961), Appendix III.
7. Gabriel Jackson, *The Spanish Republic and the Civil War* (Princeton, 1965), pp. 347-348.
8. Hugh Thomas, *The Spanish Civil War,* p. 296.
9. Alvah Bessie, *Men in Battle: A Story of Americans in Spain* (New York, 1939), p. 349.
10. *Ibid.,* p. 181.
11. Vincent Brome, *The International Brigades: Spain, 1936-39* (New York, 1966), p. 293.
12. Thomas, *Spanish Civil War,* p. 298.
13. Cedric Salter in *Try-Out in Spain* (New York, 1942), p. xiv, says that the Brigades suffered half the casualties in all battles in which they were committed.
14. Bolin, *Spain,* p. 215; Thomas, *Spanish Civil War,* p. 622; Arthur H. Landis, *The Abraham Lincoln Brigade* (New York, 1967), p. xiii.
15. J. Alvarez del Vayo, *Freedom's Battle,* trans. Eileen E. Brooke (New York, 1940), pp. 55-57; Thomas J. Hamilton, *Appeasement's Child: The Franco Regime in Spain* (New York, 1943), p. 245.
16. B. H. Liddell Hart, *The Other Side of the Hill* (London, 1951), p. 122.
17. Thomas, *Spanish Civil War,* p. 316.
18. *Ibid.,* p. 566.
19. Cecil Eby, *Between the Bullet and the Lie: American Volunteers in the Spanish Civil War* (New York, 1969), p. 119.
20. Thomas, *Spanish Civil War,* p. 519.

196

21. John H. Muste, *Say That We Saw Spain Die: Literary Consequences of the Spanish Civil War* (Seattle, 1966), pp. 32-33.

22. For an attempt to link the lineage of the Bundeswehr with the Condor Legion, see Hans Richter, "Zur Forsetzung der verbrecherischen 'Legion-Condor' Tradition in der Bundeswehr," *Interbrigadisten,* ed. Militärakademie "Friedrich Engels" (Dresden, 1966), pp. 301-309; and, in the same work, an analysis of the elitist function of the Condor Legion in Bernard Nowak, "Die 'Legion-Condor' Propaganda als der psychologischen Aufrustung in faschistischen Deutschland," pp. 283-289.

CHAPTER IV
"MOBS FOR JOBS": THE DECLINE OF SOLDIERLY HONOR

1. Winston S. Churchill, *Their Finest Hour* (New York, 1949), p. 246.
2. *Ibid.,* p. 166.
3. For surveys of Commando development, see British Ministry of Information, *Combined Operations: The Official Story of the Commandos* (New York, 1943); Peter Young, *Commando* (New York, 1969); Herbert Molloy Mason, Jr., *The Commandos* (New York, 1966); Hilary St. George Saunders, *The Green Beret* (London, 1949). For an overall view of Combined Operations equipment and doctrine, see Bernard Fergusson, *The Watery Maze: A Story of Combined Operations* (New York, 1961).
4. Albert N. Garland and Howard McGaw Smith, *Sicily and the Surrender of Italy* (Washington, D.C., 1965), pp. 95-96, 100, 253.
5. Cornelius Ryan, *The Longest Day* (New York, 1959), pp. 236-239.
6. Robert D. Burhans, *The First Special Service Force: A War History of the North Americans, 1942-44* (Washington, D.C., 1947); Robert H. Adleman and George Walton, *The Devil's Brigade* (New York, 1966).
7. Stanley W. Dziuban, *The United States Army in World War II: Military Relations Between the United States and Canada, 1939-45* (Washington, D.C., 1959), pp. 267-268.
8. *Ibid.,* p. 265.
9. For a brief survey of the Desert "Mobs for Jobs" see Arthur Swinson, *The Raiders: Desert Strike Force* (New York, 1968).
10. Virginia Cowles, *Who Dares, Wins* (New York, 1959).
11. Vladimir Peniakoff, *Popski's Private Army* (New York, 1950).
12. For a *précis* of the various assaults on the U.S. Marine Corps, see R. D. Heinl, "The Cat with More Than Nine Lives," *U.S. Naval Institute Proceedings,* June 1954, 659-671.

13. Frank O. Hough, Verle E. Ludwig, and Henry I. Shaw, *Pearl Harbor to Guadalcanal: History of United States Marine Corps Operations in World War II*, Vol. I (Washington D.C., 1959), pp. 261-262; Charles L. Updegraph, Jr., *U.S. Marine Corps Special Units of World War II* (Washington, D.C., 1972).

14. Fletcher Pratt, *The Marines' War* (New York, 1948), p. 61.

15. Robert Leckie, *Strong Men Armed* (New York, 1962), p. 52.

16. *Ibid.*, p. 125.

17. *Ibid.*, pp. 117-118.

18. *Ibid.*

19. B. H. Liddell Hart, *History of the Second World War* (New York, 1971), p. 366.

20. William Slim, *Defeat Into Victory* (New York, 1961), p. 239.

21. For a general account of the Special Force in 1944, see Slim, *Defeat Into Victory*, pp. 217-244.

22. Charles F. Romanus and Riley Sunderland, *Stilwell's Command Problems*, (Washington, D.C., 1956), p. 131.

23. Charlton Ogburn, Jr., *The Marauders* (New York, 1959), p. 30.

24. *Ibid.*, p. 275.

25. Romanus and Sunderland, *Stilwell's Command Problems*, p. 230.

26. *Ibid.*, p. 242.

27. U.S. Army Historical Division, *Merrill's Marauders* (Washington, D.C., 1945), pp. 111-112. While the official history suggests a 102° fever for three days as the minimum threshold for sick call, the author was told by a Merrill's Marauders veteran, Sergeant First Class William Nolan, in 1958, that men marched with stool from dysentery trapped in their bloused boots and that a 104° fever was the minimum allowed for admission to sick call.

28. Ogburn, *Marauders*, p. 274.

29. Claire L. Chennault, *Way of a Fighter: The Memoirs of Claire Lee Chennault* (New York, 1949), pp. 102-174; Charles F. Romanus and Riley Sunderland, *Stilwell's Mission to China* (Washington, D.C., 1953), pp. 160-162.

30. Wesley Frank Craven and James Lea Cate, *The Army Air Forces in World War II*, Vol. I (Chicago, 1948), pp. 487-489.

31. *Ibid.*, p. 174.

32. Romanus and Sunderland, *Stilwell's Mission to China*, p. 113.

33. Craven and Cate, *Army Air Forces*, pp. 506-507.

34. Otto Heilbrunn, *Warfare in the Enemy's Rear* (New York, 1963), p. 45.

35. For a detailed history, see Werner Brockdorff, *Geheimskommandos des Zweiten Weltkrieges* (Munchen-Wels, 1966).

36. Charles Foley, *Commando Extraordinary* (London, 1954), p. 33.

37. John S. D. Eisenhower, *The Bitter Woods* (New York, 1969), p. 171.

38. For Skorzeny's version of the affair, see Otto Skorzeny, *Skorzeny's Secret Missions* (London, 1957).

39. For an adversary view of Skorzeny's postwar activities, see Kurt P. Tauber, *Beyond Eagle and Swastika* (Middletown, Conn., 1967), Vol. I, pp. 241, 332-362, 547-549, and Vol. II, pp. 1110-1111; Nationale Front des Demokratischen Deutschland, *Brown Book: War and Nazi Criminals in West Germany* (Leipzig, 1965), pp. 102, 546, 549. For a defense of Skorzeny's position and a denial of his role as the would-be assassin of Eisenhower, see Siegfried Westphal, *The German Army in the West* (London, 1951), p. 184.

CHAPTER V
AIRBORNE, AIRBORNE ALL THE WAY

1. Military Intelligence Division, U.S. War Department, *Enemy Airborne Forces* (Washington, D.C., 1942), p. 4.
2. I. Dontsov and P. Livotov, "Soviet Airborne Tactics," *Military Review,* October 1964, 29-33.
3. Robert A. Kilmarx, *A History of Soviet Air Power* (New York, 1962), pp. 94-136.
4. "Parachute Tactics," *The Aeroplane,* November 6, 1935, 550.
5. Claire L. Chennault, *Way of a Fighter: The Memoirs of Claire Lee Chennault* (New York, 1949), pp. 16-17.
6. Robert Jackson, *The Red Falcons: The Soviet Air Force in Action* (London 1970), p. 77.
7. For details on operations, tactics, and equipment, see Hellmuth Reinhardt, *Russian Airborne Operations* (Heidelberg, 1952); Reinhardt in "Encirclement at Yuknov," *Military Reivew,* May 1963, 61-75, suggests truth to the oft-heard legend that troops were dropped into deep snow without parachutes. For an official Russian version of World War II operations, see Rudakov, "Thirty Years of Soviet Airborne Forces," trans. LaVergne Dale, *Military Review,* June 1961, 42-44; Kilmarx, *Soviet Air Power,* pp. 193-197; Walter Schwabedissen, *The Russian Air Force in the Eyes of German Commanders* (New York, 1960), pp. 156, 257-258, 380.
8. Bernard Fergusson, *Wavell: Portrait of a Soldier* (London, 1961).
9. Edgar O'Ballance, *The Red Army* (London, 1964), p. 213.
10. Karl Alman, *Sprüng in die Hölle* (Rastatt, 1965), pp. 11-13; U.S. War Department, *Enemy Airborne Forces,* pp. 18 ff.
11. For the best detailed accounts of German airborne operations in Norway, see Cajus Bekker, *The Luftwaffe War Diaries,* ed. and trans. Frank Ziegler (London, 1967), pp. 80-113; T. K. Derry, *The Campaign in Norway* (London, 1952).
12. Bekker, *Luftwaffe War Diaries* (Garden City, N.Y., 1968), p. 113.

13. Department of the Army, *Airborne Operations: A German Appraisal* (Washington, D.C., 1951), pp. 13-17.
14. Albert Merglen, "Air Transport: A Determining Element of Success," *Military Review,* November 1958, 13.
15. Basil Collier, *The Defence of the United Kingdom* (London, 1957), p. 84.
16. *Ibid.,* pp. 123, 125, 178, 223; B. H. Liddell Hart, *The Other Side of the Hill* (London, 1951), p. 220.
17. George E. Blau, *The German Campaign in the Balkans* (Washington, D.C., 1953), p. 5.
18. The following sources contain detailed accounts of the battle for Crete: *Decisive Battles of World War II: The German View,* ed., Hans-Adolf Jacobson and Jurgen Rohwer, trans. Edward Fitzgerald (London, 1965), pp. 106-131; Blau, *German Campaign,* pp. 119-147; U.S. War Department, *Enemy Airborne Forces,* pp. 24-31, including details of training and equipment; Gavin Long, *Greece, Crete and Syria* (Canberra, 1962), pp. 197-319, the Australian perspective and the best detailed official account; I. S. O. Playfair, *The Mediterranean and Middle East,* Vol. I (London, 1956), pp. 121-151; Freiherr von der Heydte, "Die Fallschirmtruppe in zweiten Weltkrieg," *Bilanz des zweiten Weltkrieg* (Hamburg, 1953), pp. 177-198; von der Heydte, *Return to Crete,* trans. W. Stanley Ross (London, 1959); Alman, *Sprüng in die Hölle,* pp. 28-32; I. M. G. Stewart, *The Struggle for Crete* (London, 1967); Bekker, *Luftwaffe War Diaries,* pp. 184-196.
19. *Decisive Battles,* ed. Jacobson and Rohwer, p. 131.
20. Liddell Hart, *Other Side,* p. 243.
21. *Ibid.,* p. 355.
22. James M. Gavin, *Airborne Warfare* (Washington, D.C., 1948), p. 33.
23. Gordon A. Harrison, *Cross-Channel Attack* (Washington, D.C., 1951), p. 238; *The War Reports of George C. Marshall, H.H. Arnold and Ernest J. King* (Philadelphia, 1947), p. 170.
24. Liddell Hart, *Other Side,* p. 458.
25. Chester Wilmot, *The Struggle for Europe* (New York, 1952), p. 479.
26. Hugh M. Cole, *The Ardennes: Battle of the Bulge* (Washington, D.C., 1965), pp. 215-271.
27. J. R. M. Butler, *Grand Strategy,* Vol. II (London, 1957), pp. 258-259.
28. Winston S. Churchill, *Their Finest Hour* (Boston, 1949), pp. 466, 665, 761.
29. Arthur Bryant, *The Turn of the Tide* (New York, 1957), pp. 480-481.
30. George Chatterton, *The Wings of Pegasus* (London, 1962), pp. 36-38.
31. Hilary St. George Saunders, *The Red Beret* (London, 1965), pp. 9-18.
32. Churchill, *Their Finest Hour,* p. 761.
33. R. E. Urquhart, *Arnhem* (Derby, 1960); Christopher Hibbert, *The Battle of Arnhem* (London, 1962); Saunders, *Red Beret,* pp. 206-263.
34. Bernard L. Montgomery, *The Memoirs of Field Marshal Montgomery* (Cleveland, 1958), p. 266.

35. Lyman B. Kirkpatrick, *Captains Without Eyes: Intelligence Failures in World War II* (New York, 1969), p. 222.
36. Lewis H. Brereton, *The Brereton Diaries* (New York, 1946), p. 251.
37. Gavin, *Airborne Warfare*, p. viii.
38. Kent Greenfield, Robert R. Palmer, and Bell I. Wiley, *The Army Ground Forces: The Organization of Ground Combat Troops* (Washington, D.C., 1947), p. 93.
39. Martin Blumenson, *Breakout and Pursuit* (Washington, D.C., 1961), pp. 679-680.
40. Robert Ross Smith, *Triumph in the Philippines* (Washington, D.C., 1963), pp. 118-221, 341-350, 570-571, 601.
41. Robert L. Eichelberger, *Our Jungle Road to Tokyo* (New York, 1950), p. 260; Gavin, *Airborne Warfare*, pp. 137-139.
42. Albert Merglen, "Japanese Airborne Operations in World War II," *Military Review*, July 1960, p. 45.
43. *Ibid.*, pp. 49-51.
44. F. O. Miksche, *Paratroops* (New York, 1943), pp. 73 ff.
45. Gavin, *Airborne Warfare*, p.42. Also see Frederick Morgan, *Overture to Overlord* (New York, 1950), pp. 143, 152, 204.
46. Harrison, *Cross-Channel Attack*, p. 183.
47. Barney Oldfield, "Operation Eclipse," *Army*, February 1966, 40-43.
48. Brereton, *Diaries*, pp. 260-262.
49. *Ibid.*, p. 400.
50. *Ibid.*, pp. 308 ff.
51. *Ibid.*
52. Details of air landing and air support and supply operations and tactics can be found in: William Slim, *Defeat Into Victory* (New York, 1961), pp. 450-458; S. Woodburn Kirby, *The War Against Japan*, Vol. III (London, 1961), pp. 348-364; Bernard Fergusson, *The Wild Green Earth* (London, 1956); Leonard Mosley, *Gideon Goes to War* (London, 1957), pp. 165 ff; John Masters, *The Road Past Mandalay* (London, 1962); and the more contemporary account, Lowell Thomas, *Back to Mandalay* (New York, 1951).
53. Matthew B. Ridgway, *Soldier: The Memoirs of Matthew B. Ridgway* (New York, 1956), p.54.
54. Weygand, *Histoire L'Armée Française* (Paris: Academie Française, 1961), p. 475.
55. Marcel Vigneras, *Rearming the French* (Washington, D.C., 1957), p. 1.
56. *Ibid.*, pp. 211-212, 248; Weygand, *Histoire*, pp. 475-476.
57. The account of Dienbienphu is drawn from Jules Roy, *The Battle of Dienbienphu*, trans. Robert Baldick (New York, 1965); and Bernard Fall, *Hell in a Very Small Place* (Philadelphia, 1967).
58. Weygand, *Histoire*, pp. 477 ff.

59. *Ibid.*, pp. 478-479; Paul-Marie de la Gorce, *The French Army*, trans. Kenneth Douglas (New York, 1963), p.482.
60. De la Gorce, *French Army*, pp. 480 ff.
61. Edgar S. Furniss, Jr., *De Gaulle and the French Army* (New York, 1964), p. 38.
62. De la Gorce, *French Army*, p. 480.
63. Furniss, *De Gaulle*, p. 111.
64. Otto Heilbrunn, *Conventional Warfare in the Nuclear Age* (London, 1965), pp. 103, 120; M. Antrup, "The Future of Airborne Forces," *Military Review*, September 1964, pp. 60-61; Roger F. Hardenne, "Airborne Forces in Nuclear War," *Military Review*, January 1964, 53-57.
65. Jackson, *Red Falcons*, p. 207.
66. Heilbrunn, *Conventional Warfare*, p. 120; Hardenne, "Airborne Forces," 53.
67. Gregory Blaxland, *The Regiments Depart: A History of the British Army, 1945-1970* (London, 1971).

CHAPTER VI
CYBERNETIC ELITES: THE MERGING OF MEN AND MACHINES

1. For a definition and discussion, see B. L. M. Chapman, "The Teaching of Cybernetics," *Progress in Cybernetics*, Vol. II, ed. Norbert Wiener and J. P. Schade (Amsterdam, 1965), p.201.
2. Norbert Wiener, *Cybernetics: Or Communication in the Animal and the Machine* (Cambridge, 1961).
3. For an example of ascending optimism, see Herbert A. Simon, *The New Science of Management Decision* (New York, 1960).
4. Charles Webster and Noble Frankland, *The Strategic Air Offensive Against Germany*, Vol. I (London, 1961), p. 216.
5. Constance Babington-Smith, *Air Spy* (New York, 1957), p. 96.
6. Webster and Frankland, *Strategic Air Offensive*, p. 247.
7. Michael Foot, *SOE in France* (London, 1966), p. 91.
8. Lionel F. Ellis, *Victory in the West*, Vol. I (London, 1962), p. 104.
9. J. J. Halley, *Royal Air Force Unit Histories*, Vol. I (Brentwood, 1969), pp. 24-25, 102.
10. W. H. Tantum and E. J. Hofschmidt, *The Rise and Fall of the German Air Force* (Greenwich, Conn., 1969), pp. 44, 46, 104-108.

11. Robert Sherrod, *The History of Marine Corps Aviation in World War II* (Washington, D.C. 1952), pp. 205-206.
12. Babington-Smith, *Air Spy,* p. 151.
13. Wesley Frank Craven and James Lea Cate, *The Army Air Forces in World War II,* Vol. VI (Chicago, 1955), pp. 484, 617 ff.
14. Arthur Harris, *Bomber Offensive* (New York, 1947), pp. 77-79.
15. *Ibid.,* pp. 128 ff; Noble Frankland, *Bomber Offensive* (New York, 1970), p. 49.
16. Ian Cameron, *Wings of the Morning: The Story of the Fleet Air Arm in the Second World War* (London, 1962), p. 141.
17. *The Goebbels Diaries,* ed. and trans. Louis P. Lochner (New York, 1948), p. 608.
18. For a detailed description of PFF tactics, see Denis Richards and Hilary St. George Saunders, *Royal Air Force,* Vol. II (London, 1954), pp. 150-156; and Harris, *Bomber Offensive,* pp. 131 ff., 167. For a pungent view of Harris's treatment of the PFF, see D. C. T. Bennett, *Pathfinder* (London, 1958).
19. R. J. T. Hills, *Phantom Was There* (London, 1951).
20. Harold E. Hatt, *Cybernetics and the Image of Man* (Nashville, 1968), p. 9.
21. John Marriott, "No Break in the Cold War," *New Scientist,* March 2, 1972, 466-467.
22. For example, see Ronald Clark, *The Rise of the Boffins* (London, 1967); Solly Zuckerman, *Scientists and War* (New York, 1967); C. P. Snow, *Science and Government* (Cambridge, Mass., 1961).
23. Norman Moss, *Men Who Play God* (New York, 1968), p. 119; Curtis E. LeMay and MacKinley Kantor, *Mission with LeMay* (New York, 1965), p. 473.
24. Richard G. Hubler, *SAC: The Strategic Air Command* (New York, 1958).
25. W. J. Holmes, *Undersea Victory: The Influence of Submarine Operations on the War in the Pacific* (Garden City, 1966), pp. 36-37.
26. For example, see Hans Speidel, *Invasion 1944* (Chicago, 1951); Cornelius Ryan, *The Longest Day* (New York, 1959).
27. Arthur Bryant, *Triumph in the West* (Garden City, 1959), p. 101; B. H. Liddell Hart, "The Inter-War Years, 1919-1939," *History of the British Army,* ed. Peter Young and J.P. Lawford (New York, 1970), p. 252.
28. Bernard Fergusson, *The Watery Maze* (New York, 1961), p. 299; R. Stuart Macrae, *Winston Churchill's Toyshop* (Kineton, 1971), p. 146.
29. Ellis, *Victory in the West,* p. 543.
30. R. W. Thompson, *At Whatever Cost: The Story of the Dieppe Raid* (New York, 1957), p. 190; Fergusson, *Watery Maze,* p. 336.
31. Chester Wilmot, *The Struggle for Europe* (New York, 1952), p. 265.

32. K. J. Macksey, *Armoured Crusader: A Biography of Major-General Sir Percy Hobart* (London, 1967), pp. 234 ff; Liddell Hart, *Tanks,* Vol. II, p. 332.

33. Luigi Durand de la Penne and Virgilio Spigai, "The Italian Attack on the Alexandria Naval Base," *U.S. Naval Institute Proceedings,* February 1956, 125-135.

34. For details on the 10th MAS, see Marc'Antonio Bragadin, *The Italian Navy in World War II,* trans. Gale Hoffman (Annapolis, 1957), pp. 274 ff, and J. Valerio Borghese, *Sea Devils,* trans. James Cleugh (London, 1952).

35. For a detailed example, that of Okinawa, see Samuel Eliot Morison, *Victory in the Pacific, 1945* (Boston, 1960), pp. 29-30, 120-121.

CHAPTER VII
IMAGE AND IDEOLOGY: ILLUSIONS OF ELITISM

1. Edward Meade Earle, "Lenin, Trotsky, Stalin: Soviet Concepts of War," *Makers of Modern Strategy,* ed. Edward Meade Earle (Princeton, 1944), pp. 340-347; and Sigmund Neumann, "Engels and Marx: Military Concepts of the Social Revolutionaries," *Makers of Modern Strategy,* ed. Earle, p. 170.

2. Raymond L. Garthoff, *Soviet Military Policy* (New York, 1966), pp. 36-37.

3. Albert Seaton, *The Russo-German War, 1941-45* (London, 1971), p. 291.

4. Walter Kerr, *The Russian Army* (New York, 1944), p. 103.

5. Edgar O'Ballance, *The Red Army* (New York, 1964), p. 194.

6. Alexander C. George, *The Chinese Communist Army in Action* (New York, 1967), pp. 144-145.

7. *The Politics of the Red Chinese Army: A Translation of Activities of the People's Liberation Army,* ed. Chester J. Cheng (Stanford, 1966), p. 270.

8. *Ibid.,* "Preliminary Conditions Concerning the Establishment of Four Excellence Company Units and Opinions of Several Concrete Problems," p. 392.

9. John Gittings, *The Role of the Chinese Army* (London, 1967), p. 203.

10. Franz Schurmann, *Ideology and Organization in Communist China* (Berkeley, 1966), p. 426.

11. See Thomas H. Forster, *The East German Army,* trans. Antony Buzek (London, 1967), especially Chapter 12, "Training of the Cadres," pp. 175-193.

12. John Keep, "Lenin as Tactician," *Lenin, the Man, the Theorist, the Leader: A Re-Appraisal* (New York, 1967), pp. 135-158.

REFERENCES

13. Paul Avrich, *Kronstadt 1921* (Princeton, 1970), pp. 62, 239.
14. William Whitson, "The Military: Their Role in the Policy Process," *Communist China, 1949-1969: A Twenty-Year Reappraisal,* ed., Frank N. Trager and William Henderson (New York, 1970), pp. 95-122.
15. Heinz Höhne, *The Order of the Death's Head* (New York, 1970), p.28.
16. Adolf Hitler, *Mein Kampf* (New York, 1941), p. 763.
17. Hans Bucheim, "The SS—Instrument of Domination," trans. Richard Barry, *Anatomy of the SS State* (New York, 1967), pp. 323 ff.
18. For the mechanics of prewar SS selection, see Joachim C. Fest, *The Face of the Third Reich,* trans. Michael Bullock (New York, 1970), pp.120 ff.
19. Höhne, *Death's Head,* p. 144.
20. George H. Stein, *The Waffen SS* (Ithaca, N.Y., 1966), p. 211.
21. *Ibid., p. 294.*
22. Siegfried Westphal, *The German Army in the West* (London, 1951).
23. *Ibid.*
24. For a lengthy discussion of the SS in West Germany, see Kurt P. Tauber, *Beyond Eagle and Swastika,* Vols. I and II (Middletown, Conn., 1967).
25. Peter Neumann, *The Black March,* trans. Constantine Fitzgibbon (New York, 1958).
26. William L. White, *They Were Expendable* (New York, 1942).
27. Marc'Antonio Bragadin, *The Italian Navy in the Second World War* (Annapolis, 1958), and "The Royal Italian Navy's Last Victory," *U.S. Naval Institute Proceedings,* May 1957, pp. 479-487; Giuseppe Fioravanzo, "Italian Strategy in the Mediterranean, 1940-43," *U.S. Naval Institute Proceedings,* September 1958, 65-72.
28. Peter Scott, *The Battle for the Narrow Seas* (New York, 1946).
29. Robert J. Bulkley, *At Close Quarters: PT Boats in the United States Navy* (Washington, D.C., 1962), pp. 468-488; Bryan Cooper, *The Battle of the Torpedo Boats* (New York, 1970), pp. 258-288.

CHAPTER VIII
THE ULTIMATE PASSION: THE KAMIKAZE ATTACK CORPS

1. Ruth Benedict, *The Chrysanthemum and the Sword: Patterns of Japanese Culture* (Boston, 1946), pp. 38-39.
2. Bernard Milliot, *Divine Thunder* (New York, 1971), p. 226.
3. For a detailed discussion of technological aspects, see R. L. Wehrmeister, "Divine Wind over Okinawa," *U.S. Naval Institute Proceedings,* June 1957, 633-641.

4. Roger Pineau, "Spirit of the Divine Wind," *U.S. Naval Institute Proceedings,* November 1958, pp. 23-29.
5. Wehrmeister, "Divine Wind," p. 636.
6. *Ibid.,* pp. 640-641.
7. Milliot, *Divine Thunder,* p. 228.

CHAPTER IX
THE SELECTION-DESTRUCTION CYCLE

1. For a study of the endocrine functions measured in airborne training, see H. Basowitz, H. Persky, S. J. Korchin, and R. Grinker, *Anxiety and Stress* (New York, 1955).
2. *The Psychology and Physiology of Stress,* ed. Peter G. Bourne (New York, 1969), p. 234.
3. R. H. Ahrenfeldt, *Psychiatry in the British Army in the Second World War* (New York, 1958), pp. 53-55, 71-77.
4. For Arnhem statistics, see R. E. Urquhart, *Arnhem* (Derby, 1960).
5. George Chatterton, *The Wings of Pegasus* (London, 1962), p. 245.
6. Dwight D. Eisenhower, *Crusade in Europe* (New York, 1948), pp. 279-281.
7. John L. Zimmerman, *The Guadalcanal Campaign* (Washington, D.C., 1949), p. 34. See also Henry T. Shaw and Douglas T. Kane, *Isolation of Rabaul: U.S. Marine Corps Operations in World War II,* Vol. II (Washington, D.C., 1961), p. 277.
8. Albert Merglen, "Air Transport: A Determining Element of Success," *Military Review,* November 1958, 13.
9. Hal Burton, *The Ski Troops* (New York, 1971), pp. 183-184.
10. R. L. Helmbold, "Air Battles and Ground Battles—A Common Pattern?" RAND Corporation Paper #4548, January 1971.
11. Wesley Frank Craven and James Lea Cate, *The Army Air Forces in World War II,* Vol. VI (Chicago, 1949), pp. 429-430, 546-550.
12. Leona E. Tyler, *The Psychology of Human Differences* (New York, 1965), pp. 129-130.
13. Roy R. Grinker and John P. Spiegel, *Men Under Stress* (Philadelphia, 1945), p. 8.
14. Samuel A. Stouffer, Edward A. Suchman, Leland C. DeVinney, Shirley A. Star, and Robin M. Williams, Jr., *The American Soldier: Adjustment During Army Life,* Vol. I (Princeton, 1949), p. 294.
15. Saburo Sakai, Martin Caidin, and Fred Saito, *Samurai!* (New York, 1957), pp. 29 ff.
16. Department of the Army, *Army Battle Casualties and Non-Battle Deaths in World War II: Final Report 7 December 1941–31 December 1946* (Washington, D.C., 1947), p. 112.

17. *Ibid.,* p. 5.
18. *Ibid.,* pp. 7, 47-48.
19. Charles Webster and Noble Frankland, *Strategic Air Offensive,* Vol. III (London, 1961), pp. 286-287.
20. Trevor J. Constable and Raymond F. Tolliver, *Horrido! Fighter Aces of the Luftwaffe* (New York, 1968), p. 197.
21. Samuel Eliot Morison, *History of the United States Naval Operations in World War II,* Vol. IV (Boston, 1960), p. 194.
22. Karl Doenitz, *Memoirs,* trans. R. H. Stevens (New York, 1961); Cajus Bekker, *Defeat at Sea* (New York, 1955), pp. 10-11. For a discussion of German psychological-psychiatric selection in World War II, see John Laffin, *Jackboot* (London, 1966), pp. 154 ff.
23. R.M. Younger, *Australia and the Australians* (New York, 1970), pp. 633-634.
24. Gaston Bouthoul, *War* (New York, 1962), pp. 66-68.
25. For evolving concepts in this area, see P. Medawar, "Do Advances in Medicine Lead to Genetic Deterioration?" and J. Crow, "The Quality of People: Human Evolutionary Changes," *Natural Selection in Human Populations* (New York, 1971), pp. 300-320.
26. For a short account and bibliography, see Roger A. Beaumont, " 'Seek, Strike, Destroy': The Tank Destroyer Corps in World War II," *Armor,* March-April 1971, 43-46.
27. For an overview, see Roger A. Beaumont, "War in the Clouds: Mountain Troops in Modern Warfare," *An Cosantoir: The Irish Defence Journal,* 1968, 166-171.
28. F. W. Lanchester, *Aircraft in Warfare: The Dawn of the Fourth Arm* (London, 1916).

CHAPTER X
COUNTERPOISE TO MASS-MAN: THE FUTURE OF ELITISM

1. Charles G. Pritchard, "The Soviet Marines," *U.S. Naval Institute Proceedings,* March 1972, pp. 18-30.
2. Swedish Institute for Cultural Relations with Foreign Countries, *Fact Sheets on Swedish Defense* (Stockholm, 1969); "The Price of Neutrality: The Swedish Defense System," *International Defense Review,* No. 1, 1969, 56-59; Service de l'Etat-Major Général, *L'Armée Suisse* (Berne, 1966); Office of the Armed Forces Attaché, *The Swiss Army's Recruiting and Training Systems* (Washington, D.C., 1967); and *The Military Balance* (London, 1971).
3. See *American Military Thought,* ed. Walter Millis (Indianapolis, 1966).

4. Russell F. Weigley, *History of the United States Army* (New York, 1967), and *Towards an American Army* (New York, 1962).
5. Samuel P. Huntington, *The Soldier and the State: The Theory and Politics of Civil-Military Relations* (New York, 1964).
6. Emory Upton, *The Military Policy of the United States from 1775* (Washington, D.C., 1904).
7. Joseph Bernardo and Eugene H. Bacon, *American Military Policy: Its Development Since 1775* (Harrisburg, 1955); John McAuley Palmer, *Statesmanship or War* (Garden City, 1927); and *American in Arms: The Experience of the United States with Military Organization* (New Haven, 1941).
8. For opposed views, see Eugene Kinkead, *In Every War But One* (New York, 1959); and Albert Biderman, *March to Calumny: The Story of American POWs in the Korean War* (New York, 1963).
9. S. L. A. Marshall, *Men Against Fire* (New York, 1966).
10. For example, see Siegfried Westphal, *The German Army in the West* (London, 1951), p. 130; and B. H. Liddell Hart, *The Other Side of the Hill* (London, 1951), p. 427.
11. See Omar N. Bradley, *A Soldier's Story* (New York, 1951); Elliott Roosevelt, *As He Saw It* (New York, 1946); and Ralph Ingersoll, *Top Secret* (New York, 1946).
12. D. W. Brogan, *The American Character* (New York, 1956), pp. 211-212.
13. Andrew Hacker, *The End of the American Era* (New York, 1970), pp. 11-12.
14. *The Rommel Papers,* ed. B. H. Liddell Hart, trans. Paul Findlay (New York, 1953), p. 407; and J. Lawton Collins, *War in Peacetime* (Boston, 1969), pp. 318-322.
15. Moshe Dayan, "A Soldier's Verdict on Vietnam," *Sunday Telegraph,* October 16, 1966, 6.
16. *Theory and Research on the Causes of War,* ed. Dean G. Pruitt and Richard C. Snyder (Englewood Cliffs, 1969), p. 25.
17. F. O. Miksche, *Secret Forces* (Westport, Conn., 1970), p. 18.
18. Gaetano Mosca, *The Ruling Class* (New York, 1939), pp. 240-242.
19. J. H. Rose, *The Indecisiveness of Modern War* (London, 1927), p. 45.
20. "Planes Without Pilots—Coming Defense Weapon," *U.S. News and World Report,* February 28, 1972, 56-57.
21. Ruth Shonle Cavan, *Juvenile Delinquency* (Philadelphia, 1962), p. 72.
22. Robert Shellow and Derek V. Roemer, "No Heaven for Hell's Angels," *Trans-Action,* July/August 1966, 19.

Bibliography

Records in the National Archives

Arnold, Henry H. (Gen.) to George C. Marshall (Gen.). February 29, 1944. Eisenhower Library Personal Correspondence File, Marshall #1.

Brereton, Lewis H. (Gen.) Narrative Report of Operation MARKET. Eisenhower Library Personal Correspondence File, Brereton #1.

Brereton, Lewis H. (Gen.) to General Eisenhower. August 20, 1944. Eisenhower Library Personal Correspondence File, Brereton #1.

Clark, Mark W. (Lt. Gen.) to General Eisenhower. October 15, 1943. Eisenhower Library Personal Correspondence File, Clark #1.

Eisenhower, Dwight D. (Gen.) to Mark W. Clark (Commanding Gen., Fifth U.S. Army). July 12, 1943. Eisenhower Library Personal Correspondence File, Alexander #1.

Eisenhower, Dwight D. (Gen.) to George C. Marshall (Gen.). July 17, 1943. Eisenhower Library Personal Correspondence File, Marshall #1.

Eisenhower, Dwight D. (Gen.) to Lewis H. Brereton (Commanding Gen., First Allied Airborne Army). August 22, 1944. Eisenhower Library Personal Correspondence File, Brereton #1.

Eisenhower, Dwight D. (Gen.) to George C. Marshall (Gen.). May 23, 1945. Eisenhower Library Personal Correspondence File, Marshall #2.

Marshall, George C. (Gen.) to General Eisenhower. February 10, 1944. Eisenhower Library Personal Correspondence File, Marshall #1.

Taylor, Maxwell D. (Maj. Gen.) to Omar N. Bradley (Gen.). June 20, 1944. Eisenhower Library Personal Correspondence File, Bradley #1.

Miscellaneous Unpublished Materials

Airborne Operations: A German Appraisal. Washington, D.C.: Department of the Army, 1951.

Army Battle Casualties and Non-Battle Deaths in World War II—Final Report 7 December 1941—31 December 1946. Washington, D.C.: Department of the Army, 1947.

Artillery Section, The General Board. "Organization, Equipment and Tactical Employment of Tank Destroyer Units." Heidelberg: U.S. Forces European Theater, 1946.

Committee 24, Officers Advanced Course. "The Employment of Four Tank Destroyer Battalions in the E.T.O." Ft. Knox, Ky.: The Armored School, 1950.

Dunham, Emory A. *The Tank Destroyer History.* Washington, D.C.: Historical Section, Army Ground Forces, 1946.

Govan, Thomas P. (Capt.) *History of the 10th Light Division (Alpine).* Washington, D.C.: Historical Section, Army Ground Forces, 1946. (Historical Study #28.)

Lang, Ralph W. "Tank Destroyers." Ft. Knox, Ky.: The Armored School, 1947.

Lists and Supplements. Globe Militaria, Inc. Brooklyn, N.Y., 1970-71.

Office of the Armed Forces Attaché. *The Swiss Army's Recruiting and Training Systems.* Washington, D.C.: Embassy of Switzerland, 1967.

Reinhardt, Hellmuth. *Russian Airborne Operations.* Heidelberg: Historical Division, U.S. Army Europe, 1952.

Service de L'Etat-Major Général. *L'Armée Suisse.* October 1966.

Swedish Institute for Cultural Relations with Foreign Countries. "Facts on Sweden: The Swedish Defense." Stockholm, 1969.

"Tank Destroyer Units." Document #N-12472.17. U.S. Army Command and General Staff College Library, Ft. Leavenworth, Kansas.

Templeton, Kenneth S. (ed.) *10th Mountain Division: America's Ski Troops.* Chicago, 1945. (Private publication.)

Les Troupes Aéroportées Françaises. Paris, 1968. (French Army official publication.)

Monographs, Pamphlets, and Manuals

"The Allied Armies in Italy From 3rd September, 1943 to 12th December, 1944." Supplement to the *London Gazette,* June 9, 1950.

Denton, Frank H., and Phillips, Warren. "Some Patterns in the History of Violence." RAND Paper P-3609. June 1967.

Hastings, D. W.; Wright, D. G.; and Glueck, B. *Psychiatric Experiences of the Eighth Air Force:* First Year of Combat. Josiah Macy, Jr., Foundation, 1944.

Helmbold, R. L. "Air Battles and Ground Battles—A Common Pattern?" RAND Paper #4548. January 1971.

The Military Balance. London: Institute for Strategic Studies, 1971.

Military Intelligence Division. *Enemy Airborne Forces.* Washington, D.C.: War Department, 1942.

Military Intelligence Division. *German Mountain Warfare.* Washington, D.C.: War Department, 1944.

Military Intelligence Division. *Japanese Parachute Troops.* Washington, D.C.: War Department, 1945.

Peterson, A. H.; Reinhardt, G. C.; and Conger, E. E. (ed.). *Symposium on the Role of Air Power in Counter-Insurgency and Unconditional Warfare: Chindit Operations in Burma.* RAND Monograph 3654-PR. June 1963.

Reinhardt, Hellmuth. *Russian Airborne Operations.* Heidelberg: Historical Division, U.S. Army Europe, 1952.

U.S. Army Historical Section. *Merrill's Marauders.* Washington, D.C.: War Department, 1945.

211

V.C. Volunteers and Draftees: Differentiating Characteristics. RAND Monograph RM-5647. September 1968.

Official Documents and Histories

Blakeley,H.W., *The 32nd Infantry Division in World War II.* Madison, Wis.: Thirty-Second Division History Commission, n.d.

Blau, George E. *The German Campaign in the Balkans.* Washington, D.C.: Office of the Chief of Military History, 1953.

Blumenson, Martin. *Breakout and Pursuit.* Washington, D.C.: Office of the Chief of Military History, 1961.

———. *Salerno to Cassino.* Washington, D.C.: Office of the Chief of Military History, 1969.

British Ministry of Information. *Combined Operations: The Official Story of the Commandos.* New York: Macmillan, 1943.

Bulkley, Robert J. *At Close Quarters: PT Boats in the United States Navy.* Washington, D.C.: Naval History Division, 1962.

Butler, J. R. M. *Grand Strategy.* Vol. II. London: H. M. Stationery Office, 1957.

Cole, Hugh M. *The Ardennes: Battle of the Bulge.* Washington, D.C.: Office of the Chief of Military History, 1965.

Collier, Basil. *The Defence of the United Kingdom.* London: H.M. Stationery Office, 1957.

Craven, Wesley Frank, and Cate, James Lea. *The Army Air Forces in World War II.* 7 vols. Chicago: University of Chicago, 1949-58.

Crew, F. A. E. *The Army Medical Services: Campaigns.* Vol. III. London: H.M. Stationery Office, 1959.

Derry, T. K. *The Campaign in Norway.* London: H.M. Stationery Office, 1952.

Documents on German Foreign Policy. Series D, Vol. III. Washington, D.C.: U.S. Government Printing Office, 1953.

Dziuban, Stanley W. *United States Army in World War II: Military Relations Between the United States and Canada, 1939-45.* Washington, D.C.: Office of the Chief of Military History, 1959.

Edmonds, James E. (Brig. Gen. Sir), and Davies, H. R. (Maj. Gen.) *History of the Great War: Military Operations, Italy, 1915-19.* London:

212

H.M. Stationery Office, 1949.

Foot, Michael R. D. *S.O.E. in France.* London: H.M. Stationery Office, 1966.

Freeman, Roger A. *The Mighty Eighth: Units, Men and Machines.* Garden City, N.Y.: Doubleday, 1970.

Garland, Albert N., and Smith, Howard McGaw. *Sicily and the Surrender of Italy.* Washington, D.C.: Office of the Chief of Military History, 1965.

Greenfield, Kent; Palmer, Robert R.; and Wiley, Bell I. *The Army Ground Forces: The Organization of Ground Combat Troops.* Washington, D.C.: Office of the Chief of Military History, 1951.

Gugeler, Russell A. *Combat Actions in Korea.* Washington, D.C.: Office of the Chief of Military History, 1970.

Harrison, Gordon A. *Cross-Channel Attack.* Washington, D.C.: Office of the Chief of Military History, 1951.

Hough, Frank O.; Ludwig, Verle E.; and Shaw, Henry I. *History of U.S. Marine Corps Operations in World War II.* Vols. I and II. Washington, D.C.: Historical Division, U.S. Marine Corps, 1958-63.

Johnstone, John H. *United States Marine Corps Parachute Units.* Washington, D.C.: Headquarters, U.S. Marine Corps, 1961.

Kirby, S. Woodburn. *The War Against Japan.* Vols. I, II, and III. London: H.M. Stationery Office, 1957-64.

Long, Gavin. *Greece, Crete and Syria.* Canberra: Australian War Memorial, 1962.

Mountain Operations (U.S. Army Field Manual 31-72). Washington, D.C.: Department of the Army, 1959.

Palmer, Robert R.; Wiley, Bell I.; and Keast, William R. *The Procurement and Training of Ground Combat Troops.* Washington, D.C.: Office of the Chief of Military History, 1948.

Phillips, N. C. *Official History of New Zealand in the Second World War.* Wellington, N.Z.: War History Branch, 1957.

Playfair, I. S. O. *The Mediterranean and Middle East.* Vol. II. London: H.M. Stationery Office, 1956.

Postan, M. M.; Hay, D.; and Scott, J. D. *Design and Development of Weapons: Studies in Government and Industrial Organization.* London: H.M. Stationery Office, 1964.

Richards, Denis, and Saunders, Hilary St. George. *Royal Air Force, 1939-45.* Vol. II. London: H.M. Stationery Office, 1954.

Romanus, Charles F., and Sunderland, Riley. *Stilwell's Mission to China.* Washington, D.C.: Office of the Chief of Military History, 1953.
————. *Stilwell's Command Problems.* Washington, D.C.: Office of the Chief of Military History, 1956.
————. *Time Runs Out in CBI.* Washington, D.C.: Office of the Chief of Military History, 1959.
Schwabedissen, Walter (Lt. Gen.). *The Russian Air Force in the Eyes of German Commanders.* (USAF Historical Study No. 175.) New York: Arno, 1960.
Shaw, Henry I., and Kane, Douglas T. *Isolation of Rabaul: U.S. Marine Corps Operations in World War II.* Vol. II. Washington, D.C.: U.S. Marine Corps, 1963.
Smith, Robert Ross. *Triumph in the Philippines.* Washington, D.C.: Office of the Chief of Military History, 1963.
Upton, Emory. *The Military Policy of the United States from 1775.* Washington, D.C.: U.S. Government Printing Office, 1904.
Vigneras, Marcel. *Rearming the French.* Washington, D.C.: Office of the Chief of Military History, 1957.
The War Reports of George C. Marshall, H. H. Arnold and Ernest J. King. Philadelphia: Lippincott, 1947.
Warren, John C. *Airborne Operations in World War II, European Theatre.* Maxwell Air Force Base: U.S. Air Force Historical Division, 1956.
Webster, Charles (Sir), and Frankland, Noble. *The Strategic Air Offensive Against Germany, 1939-1945.* Vols. I, II, and III. London: H.M. Stationery Office, 1961.
Zimmerman, John L. *The Guadalcanal Campaign.* Washington, D.C.: Historical Division, U.S. Marine Corps, 1949.

Books

Abramovitch, Raphael R. *The Soviet Revolution.* New York: International Universities Press, 1962.
Adleman, Robert H., and Walton, George. *The Devil's Brigade.* New York: Bantam, 1966.

Adshead, Robin (Maj.). *Gurkha: The Legendary Soldier.* London: Leo Cooper, 1971.

Ahrenfeldt, R. H. *Psychiatry in the British Army in the Second World War.* New York: Columbia University Press, 1958.

Alexandrov, Victor. *The Tukhachevsky Affair.* Trans. John Hewish. London: MacDonald, 1963.

Alleg, Henri. *The Question.* New York: Belmont Books, 1958.

Allon, Yigal. *The Making of Israel's Army.* New York: Universe, 1970.

————. *Shield of David: The Story of Israel's Armed Forces.* New York: Random House, 1970.

Alman, Karl. *Sprüng in die Hölle.* Rastatt, Germany: Erich Pabel, 1965.

Altieri, James. *The Spearheaders.* Indianapolis: Bobbs-Merrill, 1960.

Ambler, John Steward. *The French Army in Politics.* Columbus: Ohio State University, 1966.

Ambrose, Stephen E. *Upton and the Army.* Baton Rouge: Louisiana State University, 1964.

American Military History—1607-1958. Washington, D.C.: Department of the Army, 1959.

Amis, Kingsley. *New Maps of Hell.* New York: Ballantine, 1960.

Andrzejewski, Stanislaw. *Military Organization and Society.* London: Routledge & Kegan Paul Ltd., 1954.

Argenti, Philip P. *The Occupation of Chios by the Germans and Their Administration of the Island, 1941-44.* Cambridge: Cambridge University Press, 1966.

Arnold, Henry H. *Global Missions.* New York: Harper, 1949.

Avrich, Paul. *Kronstadt 1921.* Princeton: Princeton University Press, 1970.

Babington-Smith, Constance. *Air Spy: The Story of Photo-Intelligence in World War II.* New York: Harper, 1957.

Bajema, Carl J. (ed.) *Natural Selection in Human Populations.* New York: Wiley, 1971.

Barnard, Chester I. *The Functions of the Executive.* Cambridge, Mass.: Harvard University Press, 1958.

Barnard, Marjorie. *A History of Australia.* Sydney: Angus & Robertson, 1967.

Basowitz, H.; Persky, H.; Korchin, S. J.; and Grinker, R. *Anxiety and Stress.* New York: McGraw-Hill, 1955.

215

Bauer, C. *The Battle of Arnhem: The Betrayal Myth Refuted.* London: Hodder & Stoughton, 1966.

Baughman, E. Earl, and Welsh, George Schlager. *Personality: A Behavioral Science.* Englewood Cliffs, N.J.: Prentice-Hall, 1962.

Bean, C. E. W. *Anzac to Amiens: History of the Australian Armed Forces in World War I.* Sydney: Angus & Robertson, 1961.

Beer, Israel. *Security of Israel: Yesterday, Today and Tomorrow.* Tel Aviv: Amikam, 1966.

Bekker, Cajus. *The Luftwaffe War Diaries.* Trans. Frank Ziegler. Garden City, N.Y.: Doubleday, 1968.

Benedict, Ruth. *The Chrysanthemum and the Sword: Patterns of Japanese Culture.* Boston: Houghton Mifflin, 1946.

Bennett, D. C. T. *Pathfinder.* London: Frederic Muller, 1958.

Benoist-Mechin, Jacques. *Sixty Days That Shook the West.* Trans. Peter Wiles. New York: Putnam, 1963.

Berelson, Bernard, and Steiner, Gary A. *Human Behavior: An Inventory of Scientific Findings.* New York: Harcourt, Brace & World, 1964.

Bergiers, Jacques. *L'Espionnage Scientifique.* Paris: Hachette, 1971.

Bernardo, C. Joseph, and Bacon, Eugene H. *American Military Policy: Its Development Since 1775.* Harrisburg, Pa.: Stackpole, 1955.

Bessie, Alvah. *Men in Battle: A Story of Americans in Spain.* New York: Scribner, 1939.

Biderman, Albert D. *March to Calumny: The Story of American POWs in the Korean War.* New York: Macmillan, 1963.

Blair, Clay, Jr. *The Atomic Submarine and Admiral Rickover.* New York: Holt, 1954.

Blakeley, H. W. *The 32nd Infantry Division in World War II.* Madison, Wis.: Thirty-Second Infantry Division History Commission, n.d.

Blau, Peter, and Scott, W. Richard. *Formal Organizations: A Comparative Approach.* San Francisco: Chandler, 1962.

Blaxland, Gregory. *The Regiments Depart: A History of the British Army 1945-70.* London: William Kimber, 1971.

Blumenson, Martin. *Anzio:* The Gamble That Failed. Philadelphia: Lippincott, 1963.

⸻. *Bloody River.* Boston: Houghton Mifflin, 1970.

Böhmler, Rudolf. *Fallschirmjäger.* Bad Nauheim, Germany: Podzun, 1961.

216

Bolin, Luis. *Spain: The Vital Years*. New York: Lippincott, 1967.

Bolt, David. *Gurkhas*. New York: Delacorte, 1967.

Bonila, Frank. *The Failure of Elites*. Cambridge, Mass.: M.I.T. Press, 1970.

Borghese, J. Valerio. *Sea Devils*. Trans. James Cleugh. London: Melrose, 1952.

Bourne, Peter G. *Men, Stress and Vietnam*. Boston: Little, Brown, 1970.

Bouthoul, Gaston. *War*. New York: Walker, 1953.

Bowers, Claude G. *My Mission to Spain: Watching the Rehearsal for World War II*. New York: Simon & Schuster, 1954.

Boyle, Andrew. *Trenchard*. New York: Norton, 1962.

Bradley, Francis X., and Wood, H. Glen. *Paratrooper*. Harrisburg, Pa.: The Military Service Publishing Co., 1956.

Bradley, Omar N. *A Soldier's Story*. New York: Holt, 1951.

Bragadin, Marc'Antonio. *The Italian Navy in World War II*. Trans. Gale Hoffman. Annapolis, Md.: U.S. Naval Institute, 1957.

Bray, Charles W. *Psychology and Military Proficiency*. Princeton, N.J.: Princeton University Press, 1948.

Bredin, A. E. C. *The Happy Warriors*. Gillinghan, England: Blackmore, 1961.

Brereton, Lewis H. *The Brereton Diaries*. New York: Morrow, 1946.

Brinton, Crane. *The Anatomy of Revolution*. New York: Vintage Books, 1965.

Brockdorff, Werner. *Geheimkommandos des Zweiten Weltkrieges*. München-Wels: Welsermühl, 1967.

Brodie, Bernard. *Sea Power in the Machine Age*. Princeton, N.J.: Princeton University Press, 1941.

Brogan, D. W. *The American Character*. New York: Vintage Books, 1956.

Bromberger, Merry and Serge. *Secrets of Suez*. Trans. James Cameron. London: Pan Books, 1957.

Brome, Vincent. *The International Brigades: Spain 1936-39*. New York: Morrow, 1966.

Bryant, Arthur. *The Turn of the Tide*. Garden City, N.Y.: Doubleday, 1957.

———. *Triumph in the West*. Garden City, N.Y.: Doubleday, 1959.

Buchanan, Muriel. *The Dissolution of an Empire.* London: J. Murray, 1932.

Bunyan, James, and Fisher, H. H. *The Bolshevik Revolution, 1917-18.* Stanford, Calif.: Stanford University Press, 1934.

Burhans, Robert D. *The First Special Service Force: A War History of the North Americans, 1942-44.* Washington, D.C.: n.p., 1947.

Burton, Hal. *The Ski Troops.* New York: Simon & Schuster, 1971.

Butcher, Harry C. *My Three Years with Eisenhower.* New York: Simon & Schuster, 1946.

Cameron, Ian. *Wings of the Morning: The Story of the Fleet Air Arm in the Second World War.* London: Hodder & Stoughton, 1962.

Caplow, Theodore. *Principles of Organization.* New York: Harcourt, Brace & World, 1964.

Caras, Roger A. *Wings of Gold: The Story of United States Naval Aviation.* Philadelphia: Lippincott, 1965.

Carsten, F. L. *The Rise of Fascism.* Berkeley: University of California Press, 1967.

Carter, Ross S. *Those Devils in Baggy Pants.* New York: Appleton-Century-Crofts, 1951.

Cartier, Raymond. *La Seconde Guerre Mondiale.* Paris: Larousse-Paris Match, 1967.

Cavan, Ruth Shonle. *Juvenile Delinquency.* Philadelphia: Lippincott, 1962.

Cavenagh, Sandy. *Airborne to Suez.* London: William Kimber, 1965.

Chamberlin, William Henry. *The Russian Revolution.* 2 vols. New York: Macmillan, 1935.

Chatterton, George. *The Wings of Pegasus.* London: MacDonald, 1962.

Chennault, Claire L. *Way of a Fighter: The Memoirs of Claire Lee Chennault.* New York: Putnam, 1949.

Churchill, Winston S. *Their Finest Hour.* Boston: Houghton Mifflin, 1949.

―――. *Closing the Ring.* Boston: Houghton Mifflin, 1951.

Clark, Mark W. (Gen.) *Calculated Risk.* New York: Harper, 1950.

Clark, Ronald. *The Rise of the Boffins.* London: Phoenix House, 1962.

―――. *Tizard.* Cambridge, Mass.: M.I.T. Press, 1965.

Coen, Terence Creagh (Sir). *The Indian Political Service: A Study in Indirect Rule.* London: Chatto & Windus, 1971.

218

Collins, J. Lawton. *War in Peacetime.* Boston: Houghton Mifflin, 1969.

Constable, Trevor J., and Tollivor, Raymond F. *Horrido! Fighter Aces of the Luftwaffe.* New York: Macmillan, 1968.

Cookridge, E. H. *Set Europe Ablaze.* New York: Crowell, 1967.

Cooper, Bryan. *The Battle of the Torpedo Boats.* New York: Stein & Day, 1970.

————. *PT Boats.* New York: Ballantine, 1970.

Cowles, V. *Who Dares, Wins.* New York: Ballantine, 1959.

Critchell, Laurence. *Four Stars of Hell.* New York: Ballantine, 1968.

Crozier, Brian. *Franco: A Biographical History.* London: Eyre & Spottiswoode, 1967.

Cruttwell, C. R. M. F. *A History of the Great War.* Oxford: Oxford University Press, 1936.

Dayan, Moshe. *The Diary of the Sinai Campaign.* New York: Schocken, 1965.

De Gaulle, Charles. *The Army of the Future.* Philadelphia: Lippincott, 1941.

De Gramont, Sanche. *The Secret War.* New York: Dell, 1963.

Deighton, Len. *Bomber.* New York: Harper, 1970.

De la Gorce, Paul-Marie. *The French Army.* Trans. Kenneth Douglas. New York: Braziller, 1963.

Del Vayo, J. Alvarez. *Freedom's Battle.* Trans. Eileen E. Brooke. New York: Knopf, 1940.

De Seversky, Alexander P. *Victory Through Air Power.* New York: Simon & Schuster, 1942.

Deutsch, Karl. *The Nerves of Government.* New York: Free Press, 1966.

De Wilde, John C.; Popper, David H.; and Clark, Eunice. *Handbook of the War.* Boston: Houghton Mifflin, 1939.

Doenitz, Karl. *Memoirs.* Trans. R. H. Stevens. New York: Belmont, Books, 1961.

Dos Passos, John. *Three Soldiers.* New York: Modern Library, 1932.

Douglas, William Sholto. *Combat and Command.* New York: Simon & Schuster, 1966.

Earle, Edward Meade (ed.). *Makers of Modern Strategy.* Princeton: Princeton University Press, 1944.

Eby, Cecil. *Between the Bullet and the Lie: American Volunteers in the Spanish Civil War.* New York: Holt, Rinehart & Winston, 1969.

Edmonds, James E. (Brig. Gen. Sir) *A Short History of World War I.* New York: Greenwood Press, 1968.

Eichelberger, Robert L. (Lt. Gen.) *Our Jungle Road to Tokyo.* New York: Viking, 1950.

Eisenhower, Dwight D. *Crusade in Europe.* New York: Doubleday, 1948.

Eisenhower, John S. D. *The Bitter Woods.* New York: Putnam, 1969.

Ellis, Lionel F. (Maj.) *Victory in the West.* London: H.M. Stationery Office, 1962.

Erickson, John . *The Soviet High Command: A Military-Political History.* London: St. Martin's Press, 1962.

Evans, Geoffrey Charles (Lt. Gen. Sir). *The Johnnies.* London: Cassell, 1964.

Eyermann, Karl-Heinz. *Luftspionage.* Vols. I and II. Berlin: Deutscher Militärverlag, 1964.

Fainsod, Merle. *How Russia Is Ruled.* Cambridge, Mass.: Harvard University Press, 1963.

Fall, Bernard B. *Hell in a Very Small Place.* Philadelphia: Lippincott, 1967.

Fane, Francis Douglas, and Moore, Don. *The Naked Warriors.* New York: Appleton-Century-Crofts, 1956.

Farran, R. *Winged Dagger.* London: Collins, 1954.

Farrar-Hockley, A. H. *Student.* New York: Ballantine Books, 1973.

Fedotoff White, D. *The Growth of the Red Army.* Princeton: Princeton University Press, 1944.

Fehrenbach, J. R. *This Kind of War.* New York: Macmillan, 1963.

Fergusson, Bernard. *Beyond the Chindwin.* London: Collins, 1945.

———. *The Wild Green Earth.* London: St. James's Library, 1946.

———. *The Watery Maze: A Story of Combined Operations.* New York: Holt, Rinehart & Winston, 1961.

———. *Wavell: Portrait of a Soldier.* London: Collins, 1961.

Fest, Joachim C. *The Face of the Third Reich.* Trans. Michael Bullock. New York: Pantheon, 1970.

Finer, Herman. *Mussolini's Italy.* Hamden, Conn.: Archon, 1964.

Fischer, Louis. *The Life of Lenin.* New York: Harper Colophon, 1964.

Foerster, Heinz von; White, John D.; Peterson, Larry J.; and Russell, John K. (eds.) *Purposive Systems.* New York: Spartan Books, 1968.

Foley, Charles. *Commando Extraordinary.* London: Longmans, 1954.

Forster, Thomas H. *The East German Army.* Trans. Antony Buzek. London: Allen & Unwin, 1967.

Frankland, Noble. *Bomber Offensive: The Devastation of Europe.* New York: Ballantine, 1970.

Fuller, J. F. C. *Armament and History.* New York: Scribner, 1945.

Furniss, Edgar S., Jr. *De Gaulle and the French Army.* New York: Twentieth Century Fund, 1964.

Gale, Richard. *Call to Arms.* London: Hutchinson, 1968.

Garthoff, Raymond L. *Soviet Military Policy.* New York: Praeger, 1966.

Gavin, James M. *Airborne Warfare.* Washington, D.C.: The Infantry Journal Press, 1947.

Gennep, Arnold van. *The Rites of Passage.* Trans. Monika B. Vizedom and Gabrielle L. Caffee. Chicago: University of Chicago Press, 1960.

Genovese, Santiago. *Is Peace Inevitable? Aggression, Evolution and Human Destiny.* New York: Walker, 1970.

George, Alexander C. *The Chinese Communist Army in Action.* New York: Columbia University Press, 1967.

George, Claude S., Jr. *The History of Management Thought.* Englewood Cliffs, N.J.: Prentice-Hall, 1968.

Gibson, William Carleton. *Creative Minds in Medicine.* Springfield, Ill.: Thomas, 1963.

Girardet, Raoul (ed.). *La Crise Militaire Française 1945-1962: Aspects Sociologiques et Ideologiques.* Paris: A. Colin, 1964.

Girvetz, Harry K. *Democracy and Elitism.* New York: Scribner, 1967.

Gittings, John. *The Role of the Chinese Army.* London: Oxford University Press, 1967.

Glubb, John Bagot (Lt. Gen. Sir). *The Story of the Arab Legion.* London: Hodder & Stoughton, 1948.

———. *A Soldier with the Arabs.* London: Hodder & Stoughton, 1957.

Golder, Frank Alfred (ed.). *Documents of Russian History, 1914-17.* Trans. Emanuel Aronsberg. Gloucester, Mass.: Peter Smith, 1964.

Goodie, C. B. *Strategic Air Command: A Portrait.* New York: Simon & Schuster, 1966.

Graf, H. *The Russian Navy in War and Revolution—From 1914 Up to 1918.* Munich: R. Oldenbourg, 1923.

Graham, C. A. L. (Brig. Gen.) *The History of the Indian Mountain*

Artillery. Aldershot, England: Gale & Polden, 1957.

Gregor, A. James. *The Ideology of Fascism.* New York: Free Press, 1969.

Griffith, Samuel B. *The Chinese People's Liberation Army.* New York: McGraw-Hill, 1967.

Grinker, Roy R., and Spiegel, John P. *Men Under Stress.* Philadelphia: Blakiston, 1945.

Grunberger, Richard B. *The 12-Year Reich: A Social History of Nazi Germany, 1933-45.* New York: Holt, Rinehart & Winston, 1971.

Guderian, Heinz. *Panzer Leader.* Trans. Constantine Fitzgibbon. New York: Dutton, 1952.

Hacker, Andrew. *The End of the American Era.* New York: Atheneum, 1970.

Hacklander. F. W. *Scenes from the Life of a Soldier on Active Service.* London: John Murray, 1850.

Halley, J. J. *Royal Air Force Unit Histories.* Vol. I. (n.p.) Brentwood, 1969.

Hamilton, Thomas J. *Appeasement's Child: The Franco Regime in Spain.* New York: Knopf, 1943.

Harper, Frank. *Night Climb, The Story of the Skiing 10th.* London: Longmans, Green, 1946.

Harris, Arthur. *Bomber Offensive.* New York: Macmillan, 1947.

Hatt, Harold E. *Cybernetics and the Image of Man.* Nashville, Tenn.: Abingdon Press, 1968.

Heiferman, Ron. *Flying Tigers.* New York: Ballantine, 1971.

Heilbrunn, Otto. *Warfare in the Enemy's Rear.* New York: Praeger, 1963.

———. *Conventional Warfare in the Nuclear Age.* London: Allen & Unwin, 1965.

Heiman, Grover. *Aerial Photography: The Story of Aerial Mapping and Reconnaissance.* New York: Macmillan, 1972.

Heinl, Robert Debs, Jr. *Soldiers of the Sea: The United States Marine Corps, 1775-1962.* Annapolis: U.S. Naval Institute, 1962.

Herington, John. *Air War Against Germany and Italy, 1939-43.* Canberra: Australian War Memorial, 1962.

Hezlet, Arthur (Vice Adm. Sir). *The Submarine and Sea Power.* New York: Stein & Day, 1967.

Hibbert, Christopher. *The Battle of Arnhem*. London: Macmillan, 1962.
Higham, Robin. *Armed Forces in Peacetime: Britain 1918-1940, A Case Study*. London: G. T. Foulis, 1962.
————. *The Military Intellectuals in Britain 1918-1939*. New Brunswick, N.J.: Rutgers University Press, 1966.
Hills, R. J. T. *Phantom Was There*. London: E. Arnold, 1951.
Hitch, Charles J. *Decision-Making for Defense*. Berkeley: University of California Press, 1967.
————, and McKean, Roland N. *The Economics of Defense in the Nuclear Age*. Cambridge, Mass.: M.I.T. Press, 1960.
Hitler, Adolf. *Mein Kampf*. New York: Reynal & Hitchcock, 1941.
Hoare, Mike. *Congo Mercenary*. London: Robert Hale, 1967.
Höhne, Heinz. *The Order of the Death's Head*. New York: Coward-McCann, 1970.
Holmes, W. J. *Undersea Victory: The Influence of Submarine Operations on the War in the Pacific*. Garden City, N.Y.: Doubleday, 1966.
Hoyt, Edwin P. *The Army Without a Country*. New York: Macmillian, 1967.
Hubler, Richard G. *SAC: The Strategic Air Command*. New York: Duell, Sloan & Pearce, 1958.
Hunter, Mel. *Strategic Air Command*. Garden City, N.Y.: Doubleday, 1961.
Huntington, Samuel P. *The Soldier and the State: The Theory and Politics of Civil Military Relations* New York: Vintage Books, 1964.
Huston, James. *Out of the Blue*. Lafayette, Ind.: Purdue University Press, 1972.
Hutton, J. Bernard. *Frogman Spy*. New York: McDowell, Obolensky, 1960.
————. *Commander Crabb Is Alive*. New York: Award Books, 1968.
Inchbald, Geoffrey. *Imperial Camel Corps*. London: Johnson, 1970.
Infield, Glenn B. *Unarmed and Unafraid*. New York: Macmillan, 1970.
Ingersoll, Ralph. *Top Secret*. New York: Harcourt, Brace, 1946.
International Council for Philosophy and Humanistic Studies. *The Third Reich*. New York: Praeger, 1955.
Ito, Masanori. *The End of the Imperial Japanese Navy*. Trans. Andrew Y. Kuroda. London: Weidenfeld & Nicolson, 1962.

Jackson, Gabriel. *The Spanish Republic and the Civil War.* Princeton: Princeton University Press, 1965.

Jackson, Robert. *The Red Falcons: The Soviet Air Force in Action.* London: Clifton Books, 1970.

Jackson, W. G. F. *The Battle for Italy.* New York: Harper & Row, 1967.

Jacobsen, Hans-Adolf, and Rohwer, Jurgen (eds.). *Decisive Battles of World War II: The German View.* Trans. Edward Fitzgerald. London: Andre Deutsch, 1965.

James, Harold, and Sheil-Small, Denis. *The Gurkhas.* Harrisburg, Pa.: Stackpole, 1965.

Janowitz, Morris. *The Professional Soldier.* New York: Free Press, 1960.

Jerrold, Douglas. *The Royal Naval Division.* London: Hutchinson, 1923.

Jucker, Ninetta. *Italy.* New York: Walker, 1970.

Jünger, Ernst. *The Storm of Steel.* Trans. B. Creighton. London: Chatto & Windus, 1942.

Kerensky, Alexander F. *The Catastrophe.* New York: Appleton, 1927.

Kerr, Walter. *The Russian Army.* New York: Knopf, 1944.

Kilmarx, Robert A. *A History of Soviet Air Power.* New York: Praeger, 1962.

Kinkead, Eugene. *In Every War But One.* New York: Norton, 1959.

Kirkpatrick, Lyman B., Jr. *Captains Without Eyes: Intelligence Failures in World War II.* New York: Macmillan, 1969.

Klass, Philip J. *Secret Sentries in Space.* New York: Random House, 1971.

Kolkowicz, Roman. *The Soviet Military and the Communist Party.* Princeton, N.J.: Princeton University Press, 1967.

Kriegsheim, Herbert. *Getarnt, Getauscht und doch Getreu—Die Geheimnisvollen "Brandenburger."* Berlin: Bernard & Graefe, 1958.

Laffin, John. *Jackboot.* London: Cassell, 1966.

Lanchester, F. W. *Aircraft in Warfare: The Dawn of the Fourth Arm.* London: Constable, 1916.

Landis, Arthur H. *The Abraham Lincoln Brigade.* New York: Citadel, 1967.

Lang, Kurt. *Military Sociology.* Oxford, England: Blackwell, 1965.

———. *Sociology of the Military: A Selected and Annotated Bibliography.* Chicago: Inter-University Seminar on Armed Forces & Society, 1969.

Larios, José. *Combat Over Spain.* New York: Macmillan, 1966.

Larteguy, Jean. *The Centurions.* New York: Dutton, 1962.
———. *The Praetorians.* New York: Dutton, 1964.
Lasswell, Harold D., and Lerner, Daniel (eds.). *World Revolutionary Elites: Studies in Coercive Ideological Movements.* Cambridge, Mass.: M.I.T. Press, 1965.
———. Lerner, Daniel; and Rothwell, C. Easton. *The Comparative Study of Elites.* Stanford, Calif.: Stanford University Press, 1952.
Leckie, Robert. *Strong Men Armed.* New York: Random House, 1962.
LeMay, Curtis E., and Kantor, MacKinley. *Mission with LeMay.* New York: Doubleday, 1965.
Lias, Godfrey. *Glubb's Legion.* London: Evans, 1956.
Liddell Hart, B. H. *The Real War—1914-1918.* Boston: Little, Brown, 1930.
———. *The Defense of Britain.* New York: Random House, 1939.
———. *The Other Side of the Hill.* London: Cassell, 1951.
———. (ed.) *The Rommel Papers.* Trans. Paul Findlay. New York: Harcourt, Brace, 1953.
———. *The Liddell Hart Memoirs: The Later Years.* Vol. II. New York: Putnam, 1965.
———. *History of the Second World War.* New York: Putnam, 1971.
Lochner, Louis P. (ed. and trans.) *The Goebbels Diaries.* New York: Eagle Books, 1948.
Lodwick, John. *The Filibusters: The Story of the Special Boat Service.* London: Cassell, 1947.
Lovell, Stanley P. *Of Spies and Stratagems.* New York: Permabooks, 1963.
Lucas Phillips, C. E. *The Raiders of Arakan.* London: Heinemann, 1971.
Lundwall, Sam J. *Science Fiction: What It's All About.* New York: Ace, 1971.
MacCloskey, Monro. *Alert the Fifth Force.* New York: Richards Rosen Press, 1969.
MacGregor-Hasti, Roy. *The Day of the Lion: The Life and Death of Fascist Italy 1922-45.* London: MacDonald, 1963.
Macintyre, Donald (Capt.). *Fighting Under the Sea.* New York: Norton, 1965.
MacKay, Ian. *Australians in Vietnam.* Adelaide: Rigby Ltd., 1968.

Macksey, K. J. *Armoured Crusader: A Biography of Major-General Sir Percy Hobart.* London: Hutchinson, 1967.

Macrae, R. Stuart. *Winston Churchill's Toyshop.* Kineton: Roundwood Press, 1971.

Marder, Arthur J. *The Anatomy of British Seapower.* Hamden, Conn.: Archon, 1964.

Mares, William. *The Marine Machine.* Garden City, N.Y.: Doubleday, 1971.

Marshall, S. L. A. *The River and the Gauntlet.* New York: Time Book Division, 1953.

————. *Night Drop: The American Airborne Invasion of Normandy.* Boston: Little, Brown, 1962.

————. *Men Against Fire.* New York: Morrow, 1966.

Mason, Herbert Molloy, Jr. *The Commandos.* New York: Duell, Sloan & Pearce, 1966.

Masserman, Jules H. *The Biodynamic Roots of Human Behavior.* Springfield, Ill.: Thomas, 1968.

Masters, John. *The Road Past Mandalay.* London: Michael Joseph, 1962.

McCrocklin, J. J. *Garde D'Haiti.* Annapolis: U.S. Naval Institute, 1956.

McNeill, William Hardy. *The Greek Dilemma—War and Aftermath.* Philadelphia: Lippincott, 1947.

Meisel, J. H. *The Fall of the Republic: Military Revolt in France.* Ann Arbor: University of Michigan Press, 1962.

Mercer, Charles. *The Foreign Legion.* London: Arthur Barker, Ltd., 1964.

Merglen, Albert. *Surprise Warfare.* London: Allen & Unwin, 1968.

Miksche, F. O. *Paratroops.* New York: Random House, 1943.

————. *Secret Forces.* Westport, Conn.: Greenwood Press, 1970.

Milliot, Bernard. *Divine Thunder.* Trans. Lowell Blair. New York: McCall, 1971.

Millis, Walter (ed.). *The Forrestal Diaries.* New York: Viking, 1951.

————. (ed.) *American Military Thought.* Indianapolis: Bobbs-Merrill, 1966.

Mills, C. Wright. *The Power Elite.* New York: Oxford University Press, 1957.

Mills, George. *Franco.* New York: Macmillan, 1967.

226

Mitchell, Mairin. *The Maritime History of Russia, 848-1948.* London: Sedgwick & Jackson, 1949.

Montgomery, Bernard L. *The Memoirs of Field Marshal Montgomery.* Cleveland: World, 1958.

Montross, Lynn. *War Through the Ages.* New York: Harper, 1946.

Moore, Robert Lovell. *The Green Berets.* New York: Crown, 1965.

Moran, Lord (Charles McMoran Wilson). *The Anatomy of Courage.* Boston: Houghton Mifflin, 1967.

Morgan, Frederick. *Overture to Overlord.* Garden City, N.Y.: Doubleday, 1950.

———. *Peace and War: A Soldier's Life.* London: Hodder & Stoughton, 1961.

Morgan, William J. *The OSS and I.* New York: Norton, 1957.

Morison, Samuel Eliot. *Victory in the Pacific: 1945.* Boston: Little, Brown, 1960.

Mosca, Gaetano. *The Ruling Class.* New York: McGraw-Hill, 1939.

Moskowitz, Sam. *Explorers of the Infinite: Shapers of Science Fiction.* New York: World, 1963.

———. *Seekers of Tomorrow.* New York: Ballantine, 1967.

Mosley, Leonard. *Gideon Goes to War.* New York: Scribner, 1955.

Moss, Norman. *Men Who Play God.* New York: Harper & Row, 1968.

Mrazek, James E. *The Fall of Eben Emael.* New York: McKay, 1970.

Muhleisen, Hans-Otto. *Kreta 1941.* Freiburg, Germany: Rombach, 1968.

Muste, John M. *Say That We Saw Spain Die: Literary Consequences of the Spanish Civil War.* Seattle: University of Washington Press, 1966.

Nationale Front des demokratischen Deutschland. *Brown Book: War and Nazi Criminals in West Germany.* Leipzig: Zeit im Bild (c.1965).

Neave, Airey. *Saturday at M.I.9.* London: Hodder & Stoughton, 1969.

Neumann, Peter. *The Black March.* Trans. Constantine Fitzgibbon. New York: Bantam, 1958.

Nield, Eric. *With Pegasus in India—The Story of 153 Gurkha Parachute Battalion.* (Charity Subscription Publication.) Aldershot, England: n.p., 1971.

O'Ballance, Edgar. *The Red Army.* New York: Praeger, 1964.

———. *The Red Army of China: A Short History.* New York: Praeger, 1964.

Office of Personnel, Office of Strategic Services. *Assessment of Men.* New York: Rinehart, 1953.

Ogburn, Charlton, Jr. *The Marauders.* New York: Harper, 1956.

Osanka, Franklin Mark (ed.). *Modern Guerrilla Warfare: Fighting Communist Guerrilla Movements 1941-1961.* New York: Free Press, 1962.

Owen, Frank. *The Campaign in Burma.* London: Arrow Books, 1957.

Palmer, John McAuley. *Statesmanship or War.* Garden City, N.Y.: Doubleday, 1927.

————. *America in Arms: The Experience of the United States with Military Organization.* New Haven, Conn.: Yale University Press, 1941.

Paret, Peter. *French Revolutionary Warfare from Indochina to Algeria: The Analysis of a Political and Military Doctrine.* New York: Praeger, 1964.

Paret, Peter, and Shy, John W. *Guerrillas in the 1960s.* New York: Praeger, 1962.

Pareto, Vilfredo. *The Rise and Fall of the Elites: An Application of Theoretical Sociology.* Totowa, N.J.: Bedminster Press, 1968.

Pawle, Gerald. *The Secret War, 1939-45.* New York: Sloane, 1957.

Peniakoff, Vladimir. *Popski's Private Army.* New York: Crowell, 1950.

Perlmutter, Amos. *Military and Politics in Israel: Nation-Building and Role Expansion.* London: Frank Cass, 1969.

Perrault, Gilles. *Les Parachutistes.* Paris: Editions du Seuil, 1961.

Pia, Jack. *Nazi Regalia.* New York: Ballantine, 1971.

Pierce, Philip N. (Lt. Col.), and Hough, Frank O. (Lt. Col.). *The Compact History of the United States Marine Corps.* New York: Hawthorn, 1960.

Pitt, Barrie. *1918: The Last Act.* New York: Norton, 1962.

Popham, Hugh. *Into Wind: A History of British Naval Flying.* London: Hamish Hamilton, 1969.

The Politics of the Chinese Red Army: A Translation of the Bulletin of Activities of the People's Liberation Army. Trans. and compiled by J. Chester Cheng. Stanford, Calif.: Stanford University Press, 1966.

Powers, Francis Gary, and Gentry, Curt. *Operation Overflight.* New York: Holt, Rinehart & Winston, 1970.

Pratt, Fletcher. *The Marines' War.* New York: Sloane, 1948.

228

Preston, R. M. P. (Lt. Col. the Hon.) *The Desert Mounted Corps.* Boston: Houghton Mifflin, 1923.

Pruitt, Dean G., and Snyder, Richard C. (eds.) *Theory and Research on the Causes of War.* Englewood Cliffs, N.J.: Prentice-Hall, 1969.

Pugh, Marshall. *Frogman: Commander Crabb's Story.* New York: Scribner, 1956.

Reeve, E. G. *Validation of Selection Boards.* New York: Academic Press, 1972.

Reynolds, Clark C. *The Fast Carriers: The Forging of an Air Navy.* New York: McGraw-Hill, 1968.

Ridgway, Matthew B. *Soldier: The Memoirs of Matthew B. Ridgway.* New York: Harper, 1956.

————. *The Korean War.* Garden City, N.Y.: Doubleday, 1967.

Robertson, Terence. *Dieppe: The Shame and the Glory.* Boston: Atlantic Monthly/Little, Brown, 1962.

Robson, L. L. *The First A.I.F.: A Study of Its Recruitment.* Melbourne, Australia: Melbourne University Press, 1970.

Roosevelt, Elliott. *As He Saw It.* New York: Duell, Sloan & Pearce, 1946.

————. (ed.) *F.D.R.: His Personal Letters, 1928-45.* New York: Duell, Sloan & Pearce, 1950.

Roscoe, Theodore. *On the Seas and in the Skies.* New York: Hawthorn, 1970.

Rose, J. H. *The Indecisiveness of Modern War.* London: G. Bell, 1927.

Rosenstone, Robert A. *Crusade on the Left: The Lincoln Battalion in the Spanish Civil War.* New York: Pegasus, 1969.

Rossi, A. *The Rise of Italian Fascism.* Trans. Peter and Dorothy Watt. New York: Howard Fertig, 1966.

Roy, Jules. *The Battle of Dienbienphu.* Trans. Robert Baldick. New York: Harper & Row, 1965.

Ruge, Friedrich. *Der Seekrieg: The German Navy's Story, 1939-45.* Annapolis: U.S. Naval Institute, 1957.

Russell, Lord of Liverpool. *Return of the Swastika.* New York: McKay, 1969.

Ryan, Cornelius. *The Longest Day.* New York: Simon & Schuster, 1959.

————. *The Last Battle.* New York: Simon & Schuster, 1966.

Sakai, Saburo; Caidin, Martin; and Saito, Fred. *Samurai!* New York: Dutton, 1957.

Salter, Cedric. *Try-Out in Spain.* New York: Harper, 1943.

Sapolsky, Harvey M. *Creating the Invulnerable Deterrent: Programmatic and Bureaucratic Success in the Polaris System Development.* Cambridge, Mass.: M.I.T. Press, 1971.

Saunders, Hilary St. George. *The Green Beret.* London: Michael Joseph, 1949.

————. *The Red Beret.* London: Four Square Books, 1965.

Saunders, M. G. (ed.) *The Soviet Navy.* New York: Praeger, 1959.

Schelling, Thomas. *Arms and Influence.* New Haven: Yale University Press, 1966.

Schoenbaum, David. *Hitler's Social Revolution.* London: Weidenfeld & Nicolson, 1966.

Schumpeter, Joseph J. *The Sociology of Imperialism.* New York: Meridian Books, 1955.

Schurmann, Franz. *Ideology and Organization in Communist China.* Berkeley: University of California Press, 1966.

Scott, Peter. *The Battle of the Narrow Seas.* New York: Scribner, 1946.

Seaton, Albert. *The Russo-German War 1941-45.* London: Arthur Barker, Ltd., 1971.

Selznick, Philip. *Leadership in Administration.* Evanston, Ill.: Row, Peterson, 1957.

Seton-Watson, Christopher. *Italy From Liberalism to Fascism 1870-1945.* London: Methuen, 1967.

Shaw, W. B. K. *Long Range Desert Group: Its Work in Libya, 1940-43.* London: Collins, 1945.

Sherrod, Robert. *The History of Marine Corps Aviation in World War II.* Washington, D.C.: Combat Forces Press, 1952.

Simon, Herbert A. *The New Science of Management Decision.* New York: Harper & Row, 1960.

Skorzeny, Otto. *Skorzeny's Secret Missions.* London: Robert Hale, Ltd., 1957.

Slim, William (Field Marshal Lord). *Defeat Into Victory.* New York: McKay, 1961.

Snow, C. P. *Science and Government.* Cambridge, Mass.: Harvard University Press, 1961.

Sokolovski, V. D. (ed.) *Soviet Military Strategy.* Trans. Herbert S. Dinerstein, Leon Gouré, and Thomas W. Wolfe. Englewood Cliffs, N.J.: Prentice-Hall, 1963.

Sorokin, Pitrim. *The Sociology of Revolution.* Philadelphia: Lippincott, 1925.

Speidel, Hans. *Invasion 1944.* Chicago: Regnery, 1951.

Stainforth, Peter. *Wings of the Wind.* London: Falcon Press, 1952.

Stein, George H. *The Waffen SS.* Ithaca, N.Y.: Cornell University Press, 1966.

Stern, J. P. *Ernst Jünger.* New Haven, Conn.: Yale University Press, 1953.

Stewart, I. M. D. *The Struggle for Crete.* London: Oxford University Press, 1967.

Stouffer, Samuel A., *et al.* (eds.). *The American Soldier: Adjustment During Army Life.* Princeton, N.J.: Princeton University Press, 1949.

————.(eds.). *The American Soldier: Combat and Its Aftermath.* Princeton, N.J.: Princeton University Press, 1949.

————. (eds.). *Measurement and Prediction.* New York: Wiley, 1950.

Sukhanov, Nikolai. *The Russian Revolution, 1917.* Edit. and trans. by Joel Carmichael. London: Oxford University Press, 1955.

Swiggett, Howard. *March or Die.* New York: Putnam, 1953.

Swinson, Arthur. *The Raiders: Desert Strike Force.* New York: Ballantine, 1968.

Tantum, W. H. N., and Hoffschmidt, E. J. *The Rise and Fall of the German Air Force.* Old Greenwich, Conn.: We Inc., 1969.

Tauber, Kurt P. *Beyond Eagle and Swastika.* Vols. I and II. Middletown, Conn.: Wesleyan University Press, 1967.

Tedder, Arthur. *With Prejudice.* Boston: Little, Brown, 1966.

Tetens, T. H. *The New Germany and the Old Nazis.* New York: Random House, 1961.

Thomas, Hugh. *The Spanish Civil War.* New York: Harper & Row, 1961.

Thomas, Lowell. *Back to Mandalay.* New York: Greystone Press, 1951.

Thompson, Hunter S. *Hell's Angels.* New York: Random House, 1967.

Thompson, R. W. *At Whatever Cost: The Story of the Dieppe Raid.* New York: Coward-McCann, 1957.

Thompson, Victor. *Modern Organization.* New York: Knopf, 1965.

Tiger, Lionel. *Men in Groups.* New York: Random House, 1969.

Toynbee, Arnold J. *A Study of History: Abridgement of Volumes I-VI.* Abridged by D. C. Somervell. New York: Oxford University Press, 1953.

Trotsky, Leon. *The History of the Russian Revolution.* Vols. I, II, and III. Trans. Max Eastman. Ann Arbor: University of Michigan Press, 1932.

———. *My Life.* Gloucester, Mass.: Peter Smith, 1970.

Tucker, Robert C. *The Marxian Revolutionary Idea.* New York: Norton, 1969.

Tugwell, Maurice. *Airborne to Battle: A History of Airborne Warfare—1918-1971.* London: William Kimber, 1971.

Tyler, Leona E. *The Psychology of Human Differences.* New York: Appleton-Century-Crofts, 1965.

Überhorst, Horst (ed.). *Elite für die Diktatur.* Düsseldorf, Germany: Droste, 1969.

Urquhart, R. E. *Arnhem.* New York: Norton, 1958.

Vagts, Alfred. *Hitler's Second Army.* Washington, D.C.: Infantry Journal/Penguin, 1943.

———. *A History of Militarism: Civilian and Military.* New York: Free Press, 1959.

Vernadsky, George. *Lenin: Red Dictator.* Trans. Malcolm W. Davis. New Haven, Conn.: Yale University Press, 1931.

Verney, Peter. *The Micks: The Story of the Irish Guards.* London: Peter Davies, 1970.

von der Heydte, Baron. *Return to Crete.* London: W. Stanley Moss, 1959.

von der Porten, Edward P. *The German Navy in World War II.* New York: Crowell, 1969.

von Ludendorff, Erich. *Ludendorff's Own Story—August 1914-November 1918.* Vol. II. New York: Harper, 1919.

von Salamon, Ernst. *Der Fragebogen.* Hamburg, Germany: Rohwolt, 1951.

Waldron, T. J., and Gleeson, James. *The Frogmen.* London: Evans Bros., 1952.

Walsh, Maurice N. (ed.) *War and the Human Race.* Amsterdam: Elsevier, 1971.

Weigley, Russell F. *Towards an American Army.* New York: Columbia University Press, 1962.

———. *History of the United States Army.* New York: Macmillan, 1967.

Werstein, Irving. *Wake: The Story of a Battle.* New York: Crowell, 1964.

Westphal, Siegfried. *The German Army in the West.* London: Cassell, 1951.

Weygand, Maxime. *Histoire de L'Armée Française.* Paris: Academie Française, 1961.

Wheldon, John. *Machine Age Armies.* New York: Abelard-Schuman, 1968.

White, William L. *They Were Expendable.* New York: Harcourt, Brace, 1942.

Whiting, Charles. *Skorzeny.* New York: Ballantine, 1972.

Wiener, Norbert. *Cybernetics or Control and Communication in the Animal and the Machine.* 2nd edition. Cambridge, Mass.: M.I.T. Press, 1961.

Wilmot, Chester. *The Struggle for Europe.* New York: Harper, 1952.

Wilson, Andrew. *The Bomb and the Computer.* New York: Delacorte, 1968.

Wilson, R. D. *Cordon and Search.* Aldershot, England: Gale & Polden, 1949.

Wiskemann, Elizabeth. *Fascism in Italy: Its Development and Influence.* London: Macmillan, 1969.

Wolfgang, Marvin E., and Ferracuti, Franco. *The Subculture of Violence.* London: Social Science Paperbacks, 1967.

Woodward, David. *The Russians at Sea: A History of the Russian Navy.* New York: Praeger, 1966.

Woolf, S. J. (ed.) *European Fascism.* New York: Random House, 1968.

Yablonsky, Lewis. *The Violent Gang.* Baltimore: Penguin, 1970.

Young, Peter (Lt. Col.). *Bedouin Command.* London: William Kimber, 1956.

———. *Commando.* New York: Ballantine, 1969.

Young, Peter (Brig.), and Lawford, J. P. (Lt. Col.) (eds.) *History of the British Army.* New York: Putnam's, 1970.

Younger, R. M. *Australia and the Australians.* New York: Humanities Press, 1970.

Zotos, Stephanos. *Greece: The Struggle for Freedom.* New York: Crowell, 1967.

Zuckerman, Solly. *Scientists and War.* New York: Harper & Row, 1967.

233

Articles

Albrecht, Florence Craig. "Austro-Italian Mountain Frontiers," *National Geographic Magazine* (April 1915), 321-400.

Amery, Leo S. "Mountain Warfare," *Royal United Services Institution Journal* (November 1944), 331-335.

Andreyev, N. "Italian Operations in the Alps," trans. Joseph Dasher, *Military Review* (March 1941), 57-60.

Antrup, M. "The Future of Airborne Forces," *Military Review* (September 1964), 58-62.

Arnold, Henry H. (Gen.) "The Aerial Invasion of Burma," *National Geographic Magazine* (August 1944), 129-148.

Aronson, E., and Mills, J. "The Effects of Severity of Initiation on Liking for a Group," *Journal of Abnormal and Social Psychology* (1959), 177-181.

Barnes, Harry Elmer. "The Social Philosophy of Ludwig Gumplowicz," in *An Introduction to the History of Sociology,* ed. Harry Elmer Barnes (Chicago: University of Chicago Press, 1965), pp. 191-208.

"Benning Jumpers Share Tradition," *Army Times* (October 18, 1967), 10.

Böhmler, Rudolf. "Das War die Alte Fallschirmtrupp," in *Deutsches Soldaten Jahrbuch* (Munich: Schild Verlag, 1967), pp. 77-91.

Bragadin, Marc'Antonio (Cmdr.). "The Royal Italian Navy's Last Victory," *U.S. Naval Institute Proceedings* (May 1957), 479-487.

Browning, F. A. M. (Lt. Gen.) "Airborne Forces," *Royal United Services Institution Journal* (November 1944), 349-361.

Buchheim, Hans. "The SS: Instrument of Domination," trans. Richard Barry, in *Anatomy of the SS State* (New York: Walker, 1967), pp. 127-397.

Case, Frank B. "Airborne: The Tired Revolution," *Military Review* (August 1965), 86-94.

Chapla, B. C. "Infantry in Mountain Operations," *Military Review* (March 1948), 14-18.

Christiansen, John. "The Use of Smoke in Airborne Operations," *Military Review* (Fall 1961), 141-148.

Crawford, W. Rex. "Vilfredo Pareto: Residues, Derivations and the Elite," in *An Introduction to the History of Sociology,* ed. Harry Elmer Barnes (Chicago: University of Chicago Press, 1965), pp. 555-568.

234

Dar, E. H. "Mountain Warfare," *Military Review* (January 1964), 48-51.

Dayan, Moshe. "A Soldier's Verdict on Vietnam," *Sunday Telegraph* (October 16, 1966), 6.

Deighton, Len. "The Private Armies," in *The Sunday Times Book of Alamein and the Desert War,* ed. Derek Jewell (London: Sphere Books, 1967), pp. 130-137.

De la Penne, Luigi Durand, and Spigai, Virgilio. "The Italian Attack on the Alexandria Naval Base," *U.S. Naval Institute Proceedings* (February 1956), 125-135.

d'Montjamont,—. "The Future of Mountain Troops," *Military Review* (January 1948), 100-104 (from *Revue de Defense Nationale).*

Dontsov, I., and Livotov, P. "Soviet Airborne Tactics," *Military Review* (October 1964), 29-33.

Dunn, D. P. W. "Notes on Winter Warfare in the Mountains," *Military Review* (July 1943), 15-20.

Ellis, E. R. "Supply and Evacuation in Mountain Operations," *Military Review* (July 1944), 74-78.

Eustace, N. "The Brigade of Gurkhas," *The Army Quarterly* (January 1951), 207-217.

Evans, Wayne O. (Major), and Hansen, James E. (Lt. Col.). "Troop Performance in High Altitudes," *Army* (February 1966), 55-58.

Fioravanzo, Giuseppe. "Italian Strategy in the Mediterranean, 1940-43," *U.S. Naval Institute Proceedings* (September 1958), 65-72.

Friedman, Lawrence. "Art Versus Violence," *Arts in Society* (Spring-Summer 1971), 324-331.

"The Future of Airborne Forces," *The Aeroplane* (July 20, 1945), 59.

Galey, John H. "Bridegrooms of Death: A Profile Study of the Spanish Foreign Legion," *Journal of Contemporary History* (April 1969), 48-63.

Galluser, (Capt.). "War in High Mountains," *Military Review* (July 1942), 75-81.

Garthoff, Raymond. "The Military as a Social Force," in *The Transformation of Russian Society,* ed. Cyril E. Black (Cambridge, Mass.: Harvard University Press, 1960), pp. 234-271.

Gately, Matthew J. "Soviet Airborne Operations in World War II," *Military Review* (January 1967), 14-20.

Gavin, James. "Paratroops Over Sicily," *The Infantry Journal* (November 1945), 25-33.

Gerard, H. B., and Mathewson, G. C. "The Effects of Severity on Liking for a Group: A Replication," *Journal of Experimental Psychology* (July 1966), 278-287.

Gillis, Charles A., and Antrup, M. "The Future of Airborne Forces," *Military Review* (September 1964), 56-61.

Gorman, J. C. "Leaguering in Mountain Warfare," *Military Review* (November 1952), 87-89.

Grant, Roderick M. "They'll Be Coming Round the Mountain," *Popular Mechanics* (December 1942), 50-55, 168.

Greil, Lothar. "Verleugnete Soldaten," in *Deutscher Soldatenkalendar* (Munich: Schild, 1962), 153-160.

Hackett, John. "The Profession of Arms," *U.S. Naval Institute Proceedings* (April 1967), 153-160.

Hardenne, Roger F. "Airborne Forces in Nuclear War," *Military Review* (January 1964), 52-57.

Heasley, Morgan B. "Mountain Operations in Winter," *Military Review* (June 1952), 11-18.

Heilman, Ludwig. "Fallschirmjäger auf Sizilien," in *Deutscher Soldatenkalendar* (Munich: Schild, 1959), 56-64.

Heinl, R. D. (Lt. Col.) "The Cat with More Than Nine Lives," *U.S. Naval Institute Proceedings* (June 1954), 659-671.

Hofinger, W. "Maleme-Galatas," *Stand To* (April–June 1966), 6-8.

Horne, Alistair. "Continuing Cost of Dien Bien Phu," *Sunday Telegraph* (July 2, 1967), 2.

Howton, Hugh. "Revolutionary War Machine," *Soldier* (April 1968), 43-47.

Inoguchi, Rikihei (Capt.), and Nakajima, Tadashi (Cmdr.). "The Kamikaze Attack Corps," trans. Masataka Chimaya and Roger Pineau, *U.S. Naval Institute Proceedings* (September 1953), 933-945.

Jaenecke, Erwin. "Die Edelweiss Soldaten," in *Deutscher Soldatenkalendar* (Munich: Schild, 1958), 116-117.

Katzenbach, E. L. "The Horse Cavalry in the Twentieth Century: A Study in Policy Response," in *Public Policy: Yearbook of the Harvard University Graduate School of Public Administration* (1958), 120-149.

Keep, John. "Lenin As Tactician," in *Lenin: The Man, the Theorist, the Leader: A Reappraisal,* ed. Leonard Shapiro and Peter Reddaway (New York: Praeger, 1967), pp. 135-158.

Kidde, Gustave E. "Some Aspects of Mountain Warfare," *Military Review* (July 1942), 57-61.

Killian, Lewis M. "The Significance of Multiple Group Membership in Disaster," *American Journal of Sociology* (January 1952), 309-315.

Klapper, Joseph T. "What We Know About the Effects of Mass Communication: The Brink of Hope," *Public Opinion Quarterly* (Winter 1957-58), 464-471.

Konrad, R. (Gen. der G.B.J.) "Gebirgsjäger Besteigen den Elbrus," in *Deutscher Soldatenkalendar* (Munich: Schild, 1957), 22-29.

Kostelanetz, Richard. "One-Man Think Tank," *New York Times Sunday Magazine* (December 1, 1968), 55 ff.

Krammer, Arnold. "Germans Against Hitler: The Thaelmann Brigade," *Journal of Contemporary History* (April 1969), 65-83.

Krause, Frederick C. "Airborne Operations: Mobility for the Nuclear Age," *Military Review* (November 1959), 65-78.

Lang, Kurt. "Military Organizations," in *Handbook of Social Organizations,* ed. J. G. March (New York: Rand McNally, 1964), pp. 838-878.

Laquer, Walter. "Literature and the Historian," *Journal of Contemporary History* (April 1967), 5-14.

Lasswell, Harold D. "The Garrison State," *American Journal of Sociology* (January 1941), 455-468.

Laurie, Peter. "The Drop-Ins," *Sunday Times Magazine* (October 22, 1967), 29 ff.

Lauterbach, A. T. "Militarism in the Western World," *Journal of the History of Ideas* (March 1944), 446-478.

Little, Roger W. "Buddy Relations and Combat Performance," in *The New Military,* ed. Morris Janowitz (New York: Wiley, 1967), pp. 195-223.

Loewenberg, Peter. "The Psychohistorical Origins of the Nazi Youth Cohort," *American Historical Review* (December 1971), 1457-1502.

Marriott, John. "No Break in the Cold War," *New Scientist* (March 2, 1972), 466-467.

Martin, Harold H. "Paratrooper in the Pentagon," *Saturday Evening Post* (August 28, 1954), 22 ff.

Martin, Norman E. "Dien Bien Phu and the Future of Airborne Operations," *Military Review* (June 1956), 19-26.

Mayberry, H. T. (Brig. Gen.) "Tank Destroyer Battle Experience," *Military Review* (December 1943), 50-53.

Mazlish, Bruce. "Group Psychology and Problems of Contemporary History," *Journal of Modern History* (April 1968), 163-177.

McQuie, Robert. "Military History and Mathematical Analysis," *Military Review* (May 1970), 8-17.

Melville, C. L. "Mountain and Snow Warfare Training in Scotland in World War II," *Army Quarterly* (July 1950), 165-180.

Merglen, Albert. "Two Airborne Raids in North Vietnam," *Military Review* (April 1958), 14-20.

————. "Air Transport: A Determining Element of Success," *Military Review* (November 1958), 10-16.

————. "Japanese Airborne Operations in World War II," *Military Review* (July 1960), 45-51.

Miksche, F. O. "The Future of Airborne Operations," *Military Review* (October 1964), 34-38.

Motte, Raymond. "Mountain Troops in Modern Warfare," *Military Review* (August 1964), 39-45.

"Mountain Troops," *Life* (November 9, 1942), 58-63.

Murphy, Charles J. V. "The New Multi-Purpose U. S. Army," *Army* (July 1966), 21-23.

Nehring, Walther K. "Der Einsatz Russischer Fallschirmverbande im Kampfraum des XXIV Panzerkorps ewischen Cherkassy und Kiew am 24/25 September 1943," in *Deutsches Soldaten Jahrbuch* (Munich: Schild, 1963), 208-222.

Nelson, Paul D. "Personnel Performance Evaluation," in *Handbook of Military Institutions,* ed. Roger W. Little (Beverly Hills, Calif.: Sage Publications, 1971), pp. 91-121.

Nowak, Eberhard. "Die 'Legion-Condor'—Propaganda als Bestandteil der psychologischen Aufrüstung in faschistischen Deutschland," in *Interbrigadisten,* Militärakademie Friedrich Engels (n.p.: Deutscher Militärverlag, 1966), 283-289.

O'Brine, Jack. "I Am a Tank Destroyer Commander," *Popular Science* (October 1943), 102 ff.

Oldfield, Barney. "Operation Eclipse," *Army* (February 1966), 40-43.

"Parachute Tactics," *The Aeroplane* (November 6, 1935), 550.

Pasdermajian, Henry G. "Mountain Warfare," *The Military Engineer* (November-December 1937), 428-433.

"The 'Pincer Movement' from the Air," *The Aeroplane* (October 25, 1940), 454-455.

Pineau, Roger. "Spirit of the Divine Wind," *U.S. Naval Institute Proceedings* (November 1958), 23-29.

"Planes Without Pilots—Coming Defense Weapon," *U.S. News and World Report* (February 28, 1972), 56-57.

Powell, Marcus L. "Mountain Operations," *Military Review* (January 1953), 6-15.

"The Price of Neutrality: The Swedish Defense System," *International Defense Review* (#1, 1969), 56-59.

Pritchard, Charles G. "The Soviet Marines," *U.S. Naval Institute Proceedings* (March 1972), 18-30.

Razzell, P. E. "Social Origins of Officers in the Indian and British Home Army, 1758-1962," *British Journal of Sociology* (1963), 248-260.

Reinhardt, Hellmuth. "Encirclement at Yukhnov," *Military Review* (May 1963), 61-75.

Richardson, Robert C. "The Alpini: Italy's Mountain Troops," *Infantry Journal* (March 1928), 257-265.

Richter, Hans. "Zur Fortsetzung der verbrecherischen 'Legion-Condor' Tradition in der Bundeswehr," in *Interbrigadisten,* Militärakademie Friedrich Engels (n.p.: Deutscher Militärverlag, 1966), 301-309.

Rosen, Stephen. "Reading Our Culture Through Books," *Wall Street Journal* (September 16, 1971), 14.

Rudakov,——. "Thirty Years of Soviet Airborne Forces," trans. La-Vergne Dale, *Military Review* (June 1961), 42-44.

Scheff, Thomas J. "Decision Rules, Types of Error and Their Consequences in Medical Diagnosis," *Behavioral Science* (April 1965), 97-107.

Schmidt, Julius. "Horrido! Ich Bin Jagersmann," in *Deutscher Soldatenkalender* (Munich: Schild, 1960), 168-172.

Schuler, Emil. "Ruckzug der 20. Gebirgsarmee in Finland," in *Deutsches Soldaten Jahrbuch* (Munich: Schild, 1964), 234-244.

Stephens, John M. "The Growing Demand for Airborne Forces," *Military Review* (April 1961), 9-17.

Sturm, Ted R. "Backstage with the Thunderbirds," *Airman* (July 1972), 21-28.

"Tank Busters," *Popular Mechanics* (August 1943), 18-23.

"The Tank Killers," *Fortune* (November 1942), 116 ff.

"Tanks Can Be Destroyed," *Popular Science* (December 1941), 75-79, 220.

Teveth, Shabtai. "Dayan: The Man Behind the Eyepatch," *Sunday Telegraph* (July 16, 1972), 6-7.

Thornber, Hubert E. (Lt. Col.) "The Tank Destroyers and Their Use," *Military Review* (January 1943), 21-24.

Tugwell, M. A. J. "Future of Airborne Forces," *Army Quarterly* (July 1955), 155-158.

von der Heydte (Baron). "Der Fallschirmtruppe im zweiten Weltkrieg," in *Bilanz des zweiten Weltkrieg* (Hamburg: Gerhard, 1953), pp. 177-198.

Wallace, Anthony F. C. "Revitalization Movements," *American Anthropologist* (April 1956), 264-280.

Wehrmeister, R. L. (Lt. J.G.) "Divine Wind over Okinawa," *U.S. Naval Institute Proceedings* (June 1957), 633-641.

Whitson, William. "The Military: Their Role in the Policy Process," in *Communist China, 1949-1969: A Twenty Year Appraisal,* ed. Frank N. Trager and William Henderson (New York: New York University Press, 1970), pp. 95-122.

Wigington, Letcher. "Supply Problems of an Infantry Division in Mountain Operations," *Military Review* (May 1946), 49-55.

Works, Robert C. (Lt. Col.) "Postwar Mountain Training," *Military Review* (May 1946), 72-76.

Yokoi, Toshiyuki (Rear Adm.). "Kamikazes and the Okinawa Campaign," *U.S. Naval Institute Proceedings* (May 1954), 505-513.

INDEX

Abbeville Boys, 69
Abd-ed-Krim, 34
Abel, Rudolf, 119
Aerial reconnaissance, 118–26; U-2 incident, 118–19, 122; in World War II, 118, 120–26
Africa, 27, 73; East, 11, 55; North, 25, 45, 48, 49, 55, 57, 86, 90, 91, 95, 99, 103, 122, 140, 144, 164, 182
Aggression, 192
Air forces, 3, 12, 32, 42, 56, 60, 68, 69, 177–80; airborne, 44, 46, 61, 75, 77–112, 174–75, 185; Flying Tigers, 69–72; Para's Prayer, 77; Strategic Air Command, 127–29; kamikaze corps, 161–63, 165–68
Airborne Division (11th), 97
Airborne Division (25th), 105

Airborne Division (101st), 175
Alanbrooke, Lord, 83, 89, 90
Albania, 25, 56, 84
Aleutians, 164, 165
Algeria, 6, 11, 103, 107, 156
Allgemeine SS, 156
American Volunteer Group (Flying Tigers), 69–72
Anzacs, 55
API-PRU operation, 120–22, 125, 126
Arditi, 24–25
Armoured forces, 177, 182–83; 79th Division, 3, 140–42
Army Security Agency, 126
Arnold, Henry H., 100, 123
Artillery, 12–13, 22, 23
Assault Parachutist Battalion (1st), 103

241

Assault Vehicle Royal Engineers (AVRE), 141
Australia, 27–29, 164, 165, 167, 181
Australian Z Special Force, 46
Austria, 11, 29, 144, 170

Babington-Smith, Constance, 121
Bagnold, Ralph, 56
Barbusse, Henri, 192
Belgian Special Air Service, 45
Belgium, 82, 95
Bennett, D. C. T., 123–25
Berlin crisis, 189
Bessie, Alvah, 37
Bigeard, Marcel, 109
Black and Tans, 25
Bomber Command, 120–21, 122–24, 180
Bouthoul, Gaston, 181
Bradley, Omar, 8, 95, 100, 142, 190
Brandenburgers, 45, 57, 72–74
Brereton, Lewis H., 99–100
Brigade of Guards, 4, 9, 26
Britain, 26, 28, 56, 69, 70, 75, 103, 162, 163, 173, 174; in World War I, 16–22, 27, 28; in Spanish Civil War, 37, 40; in World War II, 40, 45–49, 55–56, 63–67, 74, 82–86, 94, 95, 98, 100, 136, 144, 158, 159, 169, 177, 180, 183; airborne, 88–94, 103–05, 111–12, 185; aerial reconnaissance, 119–25; information elites, 126–27; submarines, 138; 79th Armoured, 140–41; frogmen, 144–45; prop-

Britain (cont.)
aganda, 148–49; torpedo boats, 158–59; and Japan, 163–64
British Commandos, 5, 7, 45–49, 59, 60, 89, 174, 182, 185
British First Airborne Division, 175
British Force 136, 46
British Glider Pilot Regiment, 175
British Long Range Penetration Force, 63
Brooke, Alan, 83, 89, 90
Browning, Frederick, 89, 90, 100
Bruchmüller, Colonel, 21
Burma, 48, 55, 63, 65, 66–67, 70, 100, 101, 164, 165, 167, 170
Bushido, 162, 163
Butt Report, 121, 122

Canada, 27, 48, 49, 136, 141; 1st Special Service Force, 52–53
Canaris, Wilhelm, 72
Carlson, Evans (Carlson's Raiders), 60–62
Castries, Christian de, 105
Casual Detachment (1688th), 65
Central Intelligence Agency (CIA), 119, 131, 187
Ceylon, 164
Challe, Maurice, 111
Chamales, Tom, 8
Chennault, Claire L., 70–72, 80–81
Chiang Kai-shek, 69, 71
China, 6, 62, 69, 70; American Volunteer Group, 70–71; imagery and propaganda, 149, 151–52; and Japan, 163–64, 167, 170; Cold War, 190

Chindits, 63, 69, 101, 182
Churchill, Winston, 27, 28, 46–47, 48, 63, 89–90, 93, 141, 165
Ciano, Costanzo, 24
Cisterna ambush, 50–51
Civil War (U.S.), 11, 17, 32, 188–89
Civilian vs. military, 17, 117–18, 188–93
Clarke, Dudley, 47–48
Clausewitz, Karl von, 181
Cochran, Philip, 101
Cold War, 127–35, 138, 145, 189–90
Colonialism, 11
Combined Operations, 47–48, 52
Committee for Nuclear Disarmament, 132
Communism, 6, 10, 12, 29, 69, 75, 107, 190; in Spain, 33, 37–38, 41–42; in Cold War, 131, 133; imagery and propaganda, 149–52
Composite Unit (Provisional) (5307th), 65
Condor Legion, 33, 38, 40–42, 88
Conformity, 10
Congo, 112
Conscription, 2, 6, 10, 18, 102, 190
Cotton, Sidney, 119, 120, 123
Cousteau, Jacques, 145
Crabb, Edward, 145
Crete, 45, 48, 84–86, 90, 91, 94, 175
Cuba, 122, 129
Curtin, John, 28
Cybernetic elites, 113–47; aerial reconnaissance, 118–25; information elites, 126–27; Strategic

Cybernetic elites (cont.)
    Air Command, 127–28; frogmen, 144–46
Cyprus, 6, 112
Czechoslovakia, 111

D'Annunzio, Gabriele, 25
Darby, William O. (Darby's Rangers), 49–51
Darwin, Charles, 8
Deaths, 179–81
Denmark, 40, 82
Derry massacre, 185
Desert Air Force, 125
Devil's Brigade, 53
Diggers (Australian), 27–28
Dirlewanger Division, 155
Dominican Republic, 102, 112
Doolittle, James, 60, 164
Duclos, Jacques, 38
Dulles, John Foster, 129

Eagle Squadron, 53
East German Army, 151
ECLIPSE, 99
Edson, Merritt, 61, 62
Egypt, 84
Einheit Steilau, 74–75
Einsatzkommando, 156
Eisenhower, Dwight, 26, 74–75, 100, 129, 142, 175
Engels, Friedrich, 150
Ethiopia, 25, 56, 63, 144
Excellence Companies, 151
Fascism, 24–25, 29, 40
FIDO anti-fog burner systems, 125

Fighting 69th of New York, 53
Fighter Squadron (99th), 69
Finland, 81, 150
First Allied Airborne Army, (FAAA), 99
1st Guards Army, 150–51
Flying Tigers, 69–72
Formosa, 163, 164
France, 26, 28, 34, 126, 149, 152; in World War I, 11, 16, 18, 19, 21, 26–28; in Spanish Civil War, 34, 36; in World War II, 40, 46, 56, 82, 91, 95, 103, 105; airborne, 103–11; war in Vietnam, 105–07; war in Algeria, 107, 111; submarines, 138
Franco, Francisco, 30–31, 33–34, 36
Franklin, Benjamin, 79
Frederick, Robert, 52, 53
Freikorps, 25
French Foreign Legion, 26, 34, 53, 156, 170; Parachute Battalions, 103–05, 106, 109–11
Freyberg, Bernard, 85
Frisch, K. von, 119
Frogmen, 144–46, 187
Frunze, Mikhail, 150
Fuller, J. F. C., 23, 192

Galahad Force, 65–68
Galland, Adolf, 41
Gaulle, Charles de, 23, 106, 107, 109, 111
Gavin, James, 97, 102

Germany, 10, 26, 32, 69, 115, 163, 164, 174; in World War I, 11, 16, 19–23, 28, 29, 33, 74–75, 79; in World War II, 24, 47–51, 57, 59, 82–89, 90, 91–93, 95, 99–100, 131, 180, 187; in Spanish Civil War, 37, 38, 40–43; airborne, 80–89, 103, 177; aerial reconnaissance, 119–21; U-boats, 136–38, 180–81; frogmen, 144–45; propaganda, 149; East German elite forces, 151–52; torpedo boats, 159; armoured forces, 182–83; *See also* Nazis and Nazism; SS
Gero, Erno, 38
Gestapo, 21
Geyer, Captain, 21
GHQ Liaison Regiment, 126
Gibbs, Philip, 159
Gibraltar, 42, 144
Gliders, 101, 175
Goebbels, Joseph, 125
Goering, Hermann, 86, 88
Gottwald, Klement, 38
Greece, 25, 40, 56, 75, 83–84, 112; imagery and propaganda, 149
Green Berets, 7, 118
GREIF (Condor), 88
Grosz, George, 174
Ground forces, 4–5, 12
Guards Armored Training Group, 89
Guerrillas, 7, 11–12, 30, 75, 192
Gurkhas, 1, 3, 44, 63
Harris, Arthur, 120, 123, 124, 187

Hatt, Harold E., 126
Helicopters, 107, 111, 112
Henriques, Robert, 47
Heroes and heroism, 5–6, 10
Hierchos Lochos, 149
Himmler, Heinrich, 154, 155
Hitler, Adolf, 19, 26, 40, 42, 43, 45, 72–74, 83, 84, 86, 115, 154, 155
Hobart, P. C. S., 140–41, 142
Holland, 40, 82, 88, 91, 95, 121, 164, 177
Hong Kong, 164
Horthy, Miklos, 73
Hungary, 73

Imagery and ideology, 148–60
India, 27, 63, 101, 167; National Army, 63; 50th Parachute Brigade, 101
Indo-China, 6, 75, 103, 105–07, 156
Indonesia, 75, 98, 165
Infantry, 12, 14–23; in World War II, 15–16, 18
Infiltration, 20–23
Information elites, 125–31
Ingersoll, Ralph, 190
Intelligence of elites, 177–79
International Brigades, 3, 33, 36–38, 40
Ireland, 6, 11, 49
Iron and Steel Brigade, 151
Iron Brigade, 3
Ironside, Edmund, 83
Israel, 102, 159, 187
Italy, 26, 70; in World War II, 11, 18–20, 50–51, 55, 56, 59, 73, 90,

Italy (cont.)
91, 95, 177; Fascism in, 24–26; and Spanish Civil War, 32, 33, 36, 38, 41; frogmen, 144–46; torpedo boats, 158–59

Jackson, Stonewall, 11
Japan, 156; in World War II, 28, 60–61, 63, 65, 67, 69, 70, 72, 97–98, 101, 136–37, 159, 163–70; and China, 69; airborne, 98; aerial reconnaissance, 121; kamikaze corps, 161–62, 165–70; history, 162–64; pilot training, 179
Java, 98, 164
Jenner, Edward, 115
Jodl, Walther, 41
Johnson, Lyndon, 130, 183
Jünger, Ernst, 25, 156

Kamikaze attack corps, 161–62; 165–70
Keitel, Wilhelm, 41
Kennedy, John F., 7, 134, 157–58, 160
Kerensky, Alexander, 149–50
Keyes, Roger, 47
Khrushchev, Nikita, 131
Konev, Ivan, 38
Korean War, 6, 12, 28, 52, 105, 129, 189, 190; airborne in, 103
Kronstadt naval garrison, 149–50, 152
Kuomintang Army, 151

Lafayette Escadrille, 53
Laffargue, Captain, 21

Lanchester, F. W., 183
Laos, 105, 118
Lawrence of Arabia, 55
Leadership, 5, 173–74, 178, 183
League of Nations, 163
Lebanon, 102
Lee, Robert E., 11
Legion de Extranjeros, 33, 36–37
Leibstandarte Adolf Hitler, 155
LeMay, Curtis, 128, 133
Lenin, Nikolai, 149, 150, 152, 181
Lettish Rifles, 149
Lettow-Vorbeck, Paul von, 55
Liddel Hart, Basil, 19, 63
Lloyd George, David, 20
Lockheed Aircraft, 118
Long Range Desert Group (LRDG), 7, 55–56, 72
Long Range Reconnaissance Patrols, 185
Ludlow-Hewitt, Edgar, 119
Luftwaffe, 82–84, 86–88, 89; Squadron of Experts (JV44), 69

MacArthur, Douglas, 28, 66, 97, 121, 158
Madagascar, 48
Malaya, 63, 75
Malinovsky, Alexander, 38, 131
Mallory, Leigh, 175
Malraux, André, 38
Malta, 84, 86
Manchuria, 163–65
Manning, Frederick, 25
MARKET GARDEN, 91, 99
Marshall, George C., 49, 66
Marshall, S. L. A., 13, 189

Marx, Karl, 8, 150
"Mass man," 2
Masters, John, 67
Matteotti, Giacomo, 25
McCarthy, Joseph, 147, 157
McNair, Lesley, 94
McNamara, Robert S., 117
Merrill, Frank (Merrill's Marauders), 52, 65–68, 182
Midway, 164
Miksche, F. O., 99, 191
Military-industrial complex, 133
Militia, 188–89
Millan Astray, Jose, 33–34
Missiles, 118–19
Mitchell, William, 79, 85, 123
"Mobs for jobs," 44–46, 57, 62, 68, 76
Mölders, Werner, 41
Montgomery, Field Marshal Bernard, 8, 91, 100, 142
Morgenstern, Oskar, 139
Morocco, 11, 34, 36
Mosca, Gaetano, 191
Motoscafi Antisubmergilio (10th MAS), 144
Mountbatten, Lord Louis, 48, 52, 65
Mussolini, Benito, 19, 24, 26, 33, 73, 84, 88

Nachtjäger Geschwader 4, 69
Napoleon, 155
National Security Agency, 126
Navy, 20, 26, 85; submarines, 129, 134–39; frogmen, 144–46, 193; PT boats, 158–59

Nazis and Nazism, 26, 29; in Spanish Civil War, 40, 41; *Waffen* SS, 152, 154–56
Negroes, 55, 69
Nettoyeurs, 27
New Guinea, 28, 97, 158, 164, 165
New Zealand, 27, 56, 85
North Atlantic Treaty Organization (NATO), 138
Norway, 40, 47, 48, 52, 53, 82, 123, 144
Nuclear weapons, 111–12, 127, 128, 129–32, 190, 192; ICBMs, 128, 130, 131, 134; nuclear submarines, 129, 134–35, 138–39

Oboe system, 125
Officers, 5, 8, 17, 23–24, 28
Oka, 168
Okinawa, 161, 168, 169
Operation Barbarossa, 180
Operation Colossus, 90
Operation *Coronet,* 169
Operation MARKET GARDEN, 91, 99
Operation *Olympic,* 169
Operation Overlord, 175
Operations Centre Khartoum, 46
Ortega y Gasset, José, 2

Palestine, 6, 11, 20, 63, 112
"Para's Prayer," 77
Parachute Battalion (501st), 94
Parachute Brigade (1st), 98
Parachute Infantry (503rd), 97
Paratroops; *See* Airborne

Path Finder Force (PFF), 122–25, 187
Patton, George, 26, 50
Peniakoff, Vladimir, 56
Pentomic concept, 134
People's Land Army, 151
Perrault, Gilles, 107
Pershing, John J., 79
Phantom, 126
Philadelphia Black Horse Troop, 3
Philippines, 51, 61, 62, 75, 97, 98, 158, 164, 167, 168
Phoenix forces, 185
Poland, 40, 82
Polaris submarines, 129, 138, 139
Polish Parachute Brigade, 93
Popski's Private Army, 56–57
Porpoises, 146
Portugal, 75, 187
Powers, Francis Gary, 119
Prendergast, Guy, 56
Primo de Rivera, Miguel, 31
Private armies, 55, 62
Propaganda, 60, 148–49, 157, 159–60
Provisional Parachute Group, 94
PT boats, 158–60
Public relations, 135
Pyke, Geoffrey, 52, 53

Quebec Conference, 52, 65

Radar, 128
Rajk, Laszlo, 38
Ramcke, Major, 85
RAND Corporation, 117, 131, 177
Recondo teams, 185

Red Army, 150, 151
Red Guards, 149
Reddemann, Major, 21
Regimental Combat Team (442nd), 55
Reserve Forces Act (1955), 189
Revolution, 17, 18–19
Rhodesia, 56
Richthofen's Flying Circus, 68
Rickover, Hyman, 138
Ridgway, Matthew, 102
Rifs, 34
Roberts, Kenneth, 49
Robots, 113–14, 140, 147, 192
Roehm, Ernst, 154
Rohr, Captain, 21
Rokossovsky, Konstantin, 38
Rommel, Erwin, 45, 57, 86, 155
Roosevelt, Eleanor, 59
Roosevelt, Elliott, 190
Roosevelt, Franklin D., 60, 165
Roosevelt, James, 60
Root, Elihu, 189
Royal Air Force (RAF), 89, 119–20, 122; Path Finder Force, 68–69; "V" Force, 139
Royal Marine Commandos, 5, 59
Russia, 6, 11, 13, 25, 26, 69, 70, 75, 103; in World War II, 11, 17–20, 27, 45, 48, 81, 164, 165, 180, 183; and Spanish Civil War, 38, 41–42; airborne, 79–81, 111; U-2 incident, 118–19, 122; Cold War, 127, 129–31, 133–34, 139, 190; submarines, 138; imagery and propaganda, 149–52
SA (Sturmabteilung), 23, 154

Science; See Technology
SEALs, 144, 185
"Secret Army Organization," 111
Seeckt, Hans von, 181
Selection-destruction cycle, 171–84
Semmelweiss, Ignaz, 115
SHAEF, 95, 99–100, 175
Shepley, James, 66
Shock troops, 27, 29
Shumpeter, Joseph, 191
Sicherheitdienst, 156
Signal Intelligence, 126
Singapore, 164
Skorzeny, Otto, 73–75, 88
Slim, William, 65, 101
Solomon Islands, 165, 175
South Africa, 27
South East Asia Command, 63, 66
Soviet Union; See Russia
Soviet "white berets," 59
Spain, 6, 8, 11, 25, 56, 75, 121, 149, 187; Civil War, 30–43
Spanish-American War, 31, 189
Spanish Foreign Legion, 3, 32–34, 36–38, 156, 170
Special Air Service (SAS), 7, 45, 56, 57, 72, 103, 105
Special Attack Corps, 167
Special Forces, 7, 45–59, 63–65, 134, 160, 185, 190; statistics, 171–73; criteria for selection, 173
Special Naval Landing Force, 98
Special Service Brigade (1st), 67, 182
Special Service Force (1st), 3, 49–53

Specialized Armour Development Establishment, 142
SS (*Schutzstaffel*), 3, 21, 23–24, 26, 73, 88, 91, 156, 170; *Waffen* SS, 4, 23, 152, 154–57, 187; 18th SS Panzer Division, 155
Stahlhelm, 25
Stalin, Joseph, 26, 43, 150
Stalinfalken, 69
Standards of elites, 173–74, 177–79
Statistics, 171–73, 179–81
Status, 4, 6–7
Steiner, Felix, 23–24
Stilwell, Joseph, 65, 67, 71
Stirling, David, 56, 57
Storm troops, 22–24, 46
Stosstruppen, 21, 24
Strategic Air Command (SAC), 117, 127–35, 139, 140
Student, Kurt, 82, 84, 86, 91
Sturmabteilung (SA), 23, 25, 154
Submarines, 138, 183; nuclear, 129, 134–35, 139, 140; U-boats, 137, 180–81; in war, 137–38, 144, 145–46; Japanese, 169; casualties, 180–81
Suez, 102, 111
Sumatra, 164
Support forces, 13–14
Sweden, 6, 123, 187
Swiss Guards, 170
Switzerland, 6, 187
Syria, 11, 48

Taft, William H., 189
TALISMAN, 99
Tanks, 46, 116, 141–42, 182–83

Taylor, Maxwell, 102, 134
Technology, 9, 12, 13, 18–20, 114–16, 138, 177; cybernetic elites, 116–17, 127; nuclear, 127–30; Strategic Air Command, 127–33; submarines, 138; armoured forces, 140–42; frogmen, 144–46
Terrorism, 10–11
Thailand, 164
"Think tanks," 117, 118
Thoma, Wilhelm Ritter von, 41
Thomas, Hugh, 37
Thomas, Lowell, 55
Timor, 164
Tito, Marshal, 38
Togliatti, Palmiro, 38
Torpedo boats, 158–59
Totenkopfverbande, 156
Tragino Aquaeduct, 90
Trench fighting, 159
Trenchard, Hugh, 123
Trotsky, Leon, 150
Truscott, Lucian, 49
Tukhachevsky, Mikhail, 80, 81, 150
Tunisia, 48, 57
Turkey, 11, 18, 20, 26, 27, 29

U-boats, 136–37, 180
U-2 incident, 118–19, 122
Ulbricht, Walter, 38
Underwater Demolition Teams, 145
Uniforms and insignia, 2, 3, 34, 52, 102, 151

United States, 75; military forces, 26, 28, 188; in Spanish Civil War, 37, 40; 1st Special Service Force, 49–53; Pacific forces, 65–68; special air groups, 68–72; at war in Europe, 74–75; airborne, 91, 93–95, 97–105, 111, 112; in World War II, 91, 93–95, 97–101, 142, 175, 177; in Vietnam War, 112; U-2 incident, 118–19, 122; aerial reconnaissance, 121–22; information elites, 125–27; Cold War cybernetics, 126–35; Central Intelligence Agency, 131; submarines, 136–39; frogmen, 144–45; propaganda, 148–49; PT boats, 158–59; at war in Pacific, 163–69; Air Force standards, 178–79; deaths in war, 179–81; attitudes toward military, 188–93

U.S. Army Tank Destroyer Corps, 182–83

U.S. Eighth Air Force, 178

U.S. Eleventh Airborne Division, 98

U.S. Marine Corps, 5, 26, 59–62, 190; and Raider battalions, 7, 59–62; Parachute Regiment, 61; Provisional Parachute Battalion, 175

U.S. Rangers, 7, 49–52, 174, 182, 185

U.S. Special Forces, 59

U.S. Tenth Mountain Division, 177, 183

Universal military training, 6, 147, 189

Upton, Emory, 189

"V" Bomber Force, 132

V-Force, 46

Verfügungstruppe, 154

Vietnam War, 6, 7, 13, 28, 55, 118, 130, 133, 134, 139, 146, 174, 183, 185, 187, 190; battle for Dienbienphu, 105–06; airborne in, 112

Volksturm, 174

Volunteers, 3, 21, 22, 24, 27, 33, 37, 48, 62, 66

Von Hutier, General, 20

Waffen SS; See SS

Wallace, George, 133

War, 2, 7, 10–14, 181, 190; conventional, 12; and technology, 114–15

Wars of National Liberation, 6, 12, 191

Washington, George, 9

Waugh, Evelyn, 47

Wavell, Archibald P., 56, 81, 98

Weapons, 12, 18, 20, 22, 40, 42, 79; See also Technology

Wehrmacht, 155, 156

Wessel, Horst, 155

Wingate, Orde, 55, 63, 65, 66, 101

Wintringham, Tom, 37

World War I, 11–29, 32, 33, 55, 119, 144, 149, 159, 163, 174, 181, 183; airborne, 79; submarines, 136

World War II, 5, 8, 11–24, 26, 28, 42, 159, 174–75, 177–80, 183; Commandos in, 45–49; Rangers in, 49–52; 1st Special Forces in, 52–53; other special forces in, 55–56, 57, 59; U.S. Marines in, 59–62; Pacific special forces in, 62–63, 65–69; elite air groups in, 69–71; German elite groups in, 72–75; airborne in, 81–82, 90–91, 93–95, 97–103, 105; technological developments, 114–16; aerial reconnaissance, 120–21;

World War II (cont.) submarines in, 136–37; 79th Armoured in, 140–42; imagery and propaganda, 148–51, 158–59; PT boats, 158–59; kamikaze corps in, 161–64; in Pacific, 164–67; deaths in, 179–81

"Y" Service, 126
Yokosuka Special Naval Landing Forces, 98
Youth tribalism, 192
Yugoslavia, 40, 56, 84